AF517129

Climate Change & Himalayan Informatics

About the Editors

J Sundaresan, Head, Climate Change Informatics in CSIR- NISCAIR an autonomous institution under Council of Scientific and Industrial Research (CSIR), Ministry of Science & Technology, Govt of India. He is the Editor of Indian Journal of Geo-Marine Sciences (IJMS), a publication of CSIR. He is a PhD in Oceanography from Cochin University of Science & Technology, India. Thirty years research experience in coastal oceanography and twenty two years experience in Island studies. Associated with science education, science popularisation and ecological activities for the last 36 years. Actively associated with scientists Associations to sustain ethics, truth and scientific temper in scientific institutions. Published more than fifty publications, an examiner for doctoral thesis evaluation and member of national and international committees. Invited for key note address and invited talk in many Universities. Travelled in many countries.

Pankaj Gupta, Principal scientist in CSIR-Central Road Research Institute, an autonomous institute under Council of Scientist and Industrial Research (CSIR), Ministry of Science & Technology, Govt. of India. He is having more than 22 years research experience in the field of geology and in engineering geology. He worked in many research and consultancy projects related to slope stability and other Geotechnical issues with especial reference to mountain roads He did his Masters in Applied Geology from the University of Roorkee (now it is named as IIT Roorkee). He also did his Ph. D. from University of Roorkee. He has published 25 research papers in international/National Journal/conference/Symposium/seminar etc. He has received UGC/CSIR fellowship and best paper awards and guided many B.E. & M. Sc. Students for their thesis work. He was external examiner for the doctoral thesis evaluation, member of many professional bodies and member of Beouro of Indian Standard committee for making BIS Code. He is a reviewer for many research journals and has reviewed the papers for the Fifth International conference on case Histories in Geotechnical Engineering, New York, NY April 13-17, 2004. Travelled in many countries across the Europe.

K.M. Santosh has obtained M.Sc. degree in Oceanography from Cochin University of Science and Technology (CUSAT). His Ph.D is from Mangalore University in Bio-Science". Presently he is a Faculty in Department of Physical Oceanography, School of Marine Sciences, CUSAT. Dr. Santosh has more than 15 years of experience this includes scientific, research and teaching fields. He has worked as Remote Sensing Specialist in King Fahd University of Petroleum and Minerals (KFUPM), Kingdom of Saudi Arabia for the UNCC sponsored Project on "Oceanographic Survey in support of the Marine and Coastal Damage Assessment". Dr. Santosh had worked as an Associate Professor in Arba Minch University (AMU), Ethiopia. He has immense cruise experience covering entire Arabian Sea and Bay of Bengal several times and also from Arabian Gulf.. He is a reviewer of IJMS.. Participated in many national workshops and International conferences. Performed as a faculty in the area of Remote Sensing and GIS for the National level training programme on "Climate Change & Spatial Technology".

Ram Boojh, Programme specialist at the UNESCO Office in New Delhi, responsible for Ecological and Earth Sciences, World Heritage Biodiversity and natural heritage programmes and is also focal point for the UN Decade for Education for Sustainable Development (DESD). He has over 30 years of experience of working with the academic institutions, voluntary sector, Government and international organizations. He has a distinguished academic career with Doctorate in Ecology and recipient of many awards and honours including the Indian National Science Academy Medal presented by Mrs Indira Gandhi, then Prime Minister of India in January 1984. He has travelled widely and has been visiting fellow at many European and US universities and academic institutions. He has Published over 100 research/technical papers / popular articles and 11 books

Climate Change & Himalayan Informatics

Editors

J. Sundaresan
Pankaj Gupta
K.M. Santosh
Ram Boojh

SCIENTIFIC
PUBLISHERS (INDIA)

Published by:

Scientific Publishers (India)
5 A, New Pali Road, P.O. Box 91
Jodhpur 342 001 (India)

Branch Office
Scientific Publishers (India)
4806/24, Ansari Road, Daryaganj
New Delhi - 110 002 (India)

E-mail: info@scientificpub.com
Website: www.scientificpub.com

© 2013, Editors

All rights reserved. No part of this publication or the information contained herein may be reproduced, adapted, abridged, translated, stored in a retrieval system, computer system, photographic or other systems or transmitted in any form or by any means, electronic, mechanical, by photocopying, recording or otherwise, without written prior permission from the editors.

Disclaimer: Whereas every effort has been made to avoid errors and omissions, this publication is being sold on the understanding that neither the editors (or authors of chapters in edited volumes) nor the publishers nor the printers would be liable in any manner to any person either for an error or for an omission in this publication, or for any action to be taken on the basis of this work. Any inadvertent discrepancy noted may be brought to the attention of the publishers, for rectifying it in future editions, if published.

ISBN: 978-81-7233-846-6

Typeset: Rajesh Ojha
Printed in India

डॉ.टी.रामसामी
सचिव
Dr. T. RAMASAMI
SECRETARY

जय हिन्द

भारत सरकार
विज्ञान और प्रौद्योगिकी मंत्रालय
विज्ञान और प्रौद्योगिकी विभाग
टेक्नोलाजी भवन, नया महरौली मार्ग, नई दिल्ली-110 016
GOVERNMENT OF INDIA
MINISTRY OF SCIENCE & TECHNOLOGY
DEPARTMENT OF SCIENCE & TECHNOLOGY
Technology Bhavan, New Mehrauli Road, New Delhi-110 016

Foreword

Himalayan ecosystem has evolved over 65 million years and yet is vulnerable to changes on account of anthropogenic effect. It is the youngest mountain system which is dynamic and tectonically active and 31 of the Himalayan peaks are taller than 7600 meters. Three perennial river systems: Indus, Ganga and Bramaputra are part of this mountain system. More than 1.5 billion people in the plain, land and mountain regions in as many as six countries depend on the offerings of the Himalayan Ecosystem.

Whole Himalaya ecosystem is evolved and adapted to natural changes, the human civilization seeks revolutionary developments and forces man-made changes. Adverse changes in the Himalaya ecosystem could affect generations to come. Therefore, it is prudent to design the man-made changes within the carrying capacity of the Himalayan ecosystem. This would require a multi-dimensional decision support system for adaptation to changes and mitigating undesirable changes.

The Book "Climate Change and Himalayan Informatics" consists of many studies on climate change and urbanization. Application of geospatial technology and modeling is made. Impacts of climate change on cropping strategies by mountain communities and threat to agro biodiversity and its conservation are discussed. Paleo-climatic conditions are examined with studies and data collected from Arctic ecosystem. Institutional capacities for sustenance of Himalayan ecosystem are part of the session Himalaya informatics and database.

Probable responses of Himalayan ecosystem due to impact of climate change have been mapped providing inputs to the perspective of recent urbanization in many regions of Himalaya. The ecosystem is prone to periodic and episodic changes of tectonically active region and climate vulnerability and changes brought out by human activities. Causality of changes on the ecosystem needs to be studied, quantified and efforts should be made to counter ill impacts of anthropogenic activities. The present book is an effort in the direction of sustaining Himalayan Ecosystem. I congratulate the authors and editors for this valuable contribution with a noble interest.

(T. Ramasami)

Tel. : 011-26510068, 011-26511439 • Fax : 011-26863847, 011-26862418 • E-mail : dstsec@nic.in

Preface

Himalaya have exotic control on the process of monsoon intensification hence Climate Change. Himadri was fast growing ever since the geodynamic evolution of Himalaya during 60-45 million years. It has reached a critical height so as to deviate the west to east flow of wind. The depression developed in this region had sucked the moist winds from Indian ocean and caused heavy spells of rain hence intensified the monsoon. This phenomena generated upwelling in the selected oceanic borders of Indian subcontinent before 8.5 million years. Rainfall had caused augmented sedimentation, rich biodiversity and splendid gene pool in Himalayan region and its plains. Indo-gangetic plain evolved due to the natural process occurred at Himalaya is the life line of millions of people. This become one of the most densely populated region of the world (250 persons/km^2 in Nepal to 800 persons/km^2 in Bangladesh). According to reports related to IPCC Indo-Gangetic plain is a major emitter of the climatic forcing gases and particles viz. CO_2, CH_4, N_2O, CO, NO_x, BC and OC.

Climate change is a fact and its impact on the fragile mountain ecosystem is beyond doubt. Souring urbanization in and around mountain region and its impact is a new challenge for planning and development in this region. There are many models and experiences on mountain urbanization and its impact on ecosystem in developed countries. Whereas, studies, information and database related to the impact of Climate Change on the Himalayan ecosystem, ever increasing human settlement and urbanization are few. Present book "Climate Change and Himalayan Informatics" consists four sessions with sixteen chapters on various aspects of impact of climate change on Himalayan environment and its inhabitants.

Geospatial technology is a primary tool to develop methodology, to understand the phenomena and to prepare sensible target for adaptation and mitigation. There are four chapters in the session "Geospatial Technology, Modelling, Mitigation and Adaptation". The first chapter "Modelling energy and mass balance of Chhota Shigri Glacier" infers that the energy balance can be used as one of the approaches in estimating the mass balance of glaciers and in the recent years there has been increasing interest in predictive tools for spatially distributed estimates of mass balance further promoted by increased availability of remote sensing data. This study is the modelling of energy balance and computing the heat energy available for mass balance change in Chhota Shigri Glacier in the

Chandra Basin, Himalaya. Geospatial tools are applied to identify vulnerable zone for multipile hazards in the second chapter entiled "Natural hazards due to climate change in Himalaya and multicriteria decision analysis to identify vulnerable zone". This study infers that region of medium drainage density, higher fracture density, high degree of slope and higher relative relief will be vulnerable and is adroit to landslide. Sustainable ecosystem development planning in Phakot sub-watershed, Tehri Garhwal district, Uttarakhand using Ikonos data is the third chapter in this session. Information generated in this study would form a baseline data in future, monitoring periodic changes in the parameters of natural resources. The consequences of Climate Change – Risks to the Globe, suggest effective implementation of the United Nations Framework Convention on Climate Change (UNFCCC).

Socio-economic studies are highly beneficial for planning and development of the impact of Climate Change in any ecosystem. The session socio-economic perspectives consists six chapters. Impacts of climate change and coping strategies by mountain communities of Uttarakhand, a Himalayan states that climatic variability with in short horizontal distance due to altitudinal variation resulting in high degree of variation in vegetation and hydrological characteristics make mountain more vulnerable to any climate change phenomena. Above study infers that communities are adopting various coping strategies such as change in agricultural calendar, change in crops and crop verities, re-adoption of traditional water management systems, alternative employment and out-migration etc, to overcome the problems resulting from climate change. Vegetation of the cold deserts of Himalaya are ascribed in chapter 6 of this session. An intensive and extensive survey of the area and critical study of the plants have resulted in the documentation of 1405 species, distributed under 490 genera and 98 families of flowering plants. This is a highly fragile and vulnerable mountain ecosystem in the country. A large number of species growing in this area are of medicinal and economic value which is used by local inhabitants. The threat to agro biodiversity and its conservation in Uttarakhand Himalaya is examined in chapter 7. Gene pool of traditional crops has also been continuously enriched by gene flow across the Himalayan states and countries through trading of grain and exchange of seeds. Present study concludes that the gene pool and the traditional agriculture systems are under threat. Because of diverse ecological and climatic conditions existing in Himalaya, it harbors several ecologically and economically important plant species like *Olea ferruginea Royle*. It showed lower relative water content, lower water potential and lower water content at saturation, but a higher leaf mass per unit area, higher density of foliar tissue, higher succulence and higher chlorophyll content compared to other co-occurring species, indicating its greater tolerance to drought. Chapter 8 consists the potential tree crop for sustainable development of North-west Himalaya. The chapter 9 suggests integrating local perception and traditional knowledge for adaptation strategies in Central Himalayan agriculture from the impact of Climate Change. This study infers that very recently many traditional rural communities inhabited in different agro-climatic zones of this region are struggling through different adaptation measures as an attempt to reduce the risk of climate change vulnerability. Facebook supports numerous emergency-related organizations, including Information Systems for Crisis Response and Management (ISCRAM), Humanitarian Free and Open Source Software (FOSS) Project, as well as numerous universities with disaster-related programs. The last chapter (Chapter 10) of this session is "Climate change and social media: Facebook as an effective climate messenger".

Conservation, management, adaptation and mitigation due to the disastrous effect of Climate change is imminent possible through informatics and database of the region. The session III "Himalaya Informatics and Database" have three chapters. A permanent GPS station at Almora has been selected for the study on diurnal variation in perceptible water vapors over Almora. Ecosystem based adaptation has therefore emerged as a potential approach. Institutional mechanism for the development of technologies and their usages required for the sustenance of Himalayan ecosystem. The focus will be on promoting understanding of climate change, adaptation and mitigation, energy solutions and conserve biodiversity. Chapter 12 of this session examines the institutional capacities for sustenance of Himalayan ecosystem. The information and data related to dumping site of MSW of the Guwahati Municipal Corporation is in the fringe area of Deepar Beel, a Ramsar site known for migratory birds and aquatic biodiversity is the last chapter of this session.

Cooler conditions persists and glaciations developed in higher regime of Himalaya. The cooler regions have a profile contemporary to Europe. The last session of this book ascribes the paleoclimatic conditions and also in search of possible indicators of climate change. The chapter 14 "*Cedrus deodara* (Roxb.) G. Don a possible indicator of climate change in Eastern Himalaya" concludes that climate change posing a threat to the production and productivity of the cash crops and livestock products in Sikkim. The chapter associates with "Impact of climate change on heterotrophic bacterial communities in the water and sediment of Kongs fjord in Norwegian Arctic" (Chapter 15). It suggests that consistent isolation of mesophilic bacteria from this high arctic fjord may be an indication of increased melting of arctic ice cover. Benthic faunal assemblage diversity and time scale changes in the ice cold Kongs fjorden Arctic ecosystem, is the last chapter of this book (Chapter 16). This is a barbinger on the influence of the global climate change impacts in the arctic represented by Kongs fjorden system.

Himalaya is legendary and a unique ecosystem uplifted by the eternal geodynamic phenomena of nature. Climate and Himalaya are dual and are dialectic. The book "Climate Change and Himalayan Informatics" is an attempt of dedicated researchers who had collected invaluable data from the rugged terrains of Himalaya during hostile weather. This book consists information and data from multidisciplinary aspects of impact of climate change and on mountain ecosystem. Himalayan mountain ecosystem is in the process of rapid urbanization. Information and data to develop multispectral management and decision support system for adaptation and mitigation to this fragile ecosystem are yet to be formulated. Present book is an important initiative for the above. Support from researchers, reviewers and publishers had helped the editors to bring out this book within a short period of time. Editors acknowledge the cover photos from the contributiors of Dr. Achuta Nand Shukla & S.K. Srivastava, Botanical Survey of India and Dr. Pankaj Gupta, CSIR- Central Road Research Institute. This book will be highly beneficial to researchers, administrators and students interested in the study of Climate Change and Himalaya.

J Sundaresan
Pankaj Gupta
K.M. Santosh
Ram Boojh

Contents

Session-III
Himalaya Informatics and Database

Session-IV
Paleo Climatic Conditions

Session- I

Geospatial Technology, Modelling, Mitigation and Adaptation

Modelling energy and mass balance of Chhota Shigri glacier

Gunjan Rastogi & Ajai

Marine, Geo and Planetary Sciences Group, Space Applications Centre (ISRO), Ahmedabad-380 015, India
[E-Mail: gunjan@sac.isro.gov.in]

Introduction

Numerical glacier models quantify the sensitivity of glaciers to climate, and provide tools for evaluating the closely related response of ecosystems and water resources in glacierized regions of the world. In the complex relationship between glaciers and climate, one of the key processes is melt of snow and ice at the glacier surface. Melting followed by runoff accounts for most of the ablation on many glaciers [Anslow *et al.*, 2005]. Glacier melt models have been applied over vast regions, even globally, by calibrating the models to well-studied glaciers and applying them unchanged over much larger areas. One of the important aspects which is needed to be studied is the energy balance over the glaciers as melting and thawing of the glaciers depends largely on it. Both physically-based melt models, which are derived from a surface energy balance, and empirically-based melt models, which correlate melt with temperature and in some instances radiation, have been developed for glacierized regions. In practice physically-based melt models employ both measured and empirical parameters, while empirical melt models use empirical parameters alone [MacDougall and Flowers, 2010].

The energy balance study describes the physical connection between ice/ snow ablation and climate forcing and so it can be used as one of the approaches in estimating the mass balance of glaciers. In the recent years there has been increasing interest in predictive tools for spatially distributed estimates of mass balance further promoted by increased availability of remote sensing data. Generally, computation of snow melt is made using either an energy balance approach or using some index approach. An energy balance method requires information on radiation energy, sensible and latent heat fluxes. Several meteorological parameters are monitored to obtain this information over the snowpack. A thorough understanding of the basic energy transfer processes and their role in melting of snow packs is required [Singh and Singh (2001). Snow and Glacier Hydrology]. So in this study we have developed a physically based, surface energy balance model to simulate ablation and mass balance on Chhota Shigri glacier of Chandra basin.

Materials and methods

The Chhota Shigri glacier covering an area of 15.7 km^2, is located at a distance of about 36 km SE of Rohtang Pass. It lies on the Chandra River basin on the northern poles of the Pir Panjal range in the Lahaul and Spiti valley of Himachal Pradesh, in the western Himalaya bound by latitudes 32.19-32.28ºN and longitudes 77.49-77.55ºE. Its reported maximum elevation is 6263 m a.s.l, snout position is nearly 4050 m a.s.l., glacier length is ~ 9 km and width of about 1.5 km in its widest portion. It is included in the upper basin of the Chandra River, contributing Chenab River, one of the tributaries of the Indus river basin (Table 1). This valley-type glacier is oriented roughly north-south in its ablation area, but variety of orientations in the accumulation area (Fig. 1).

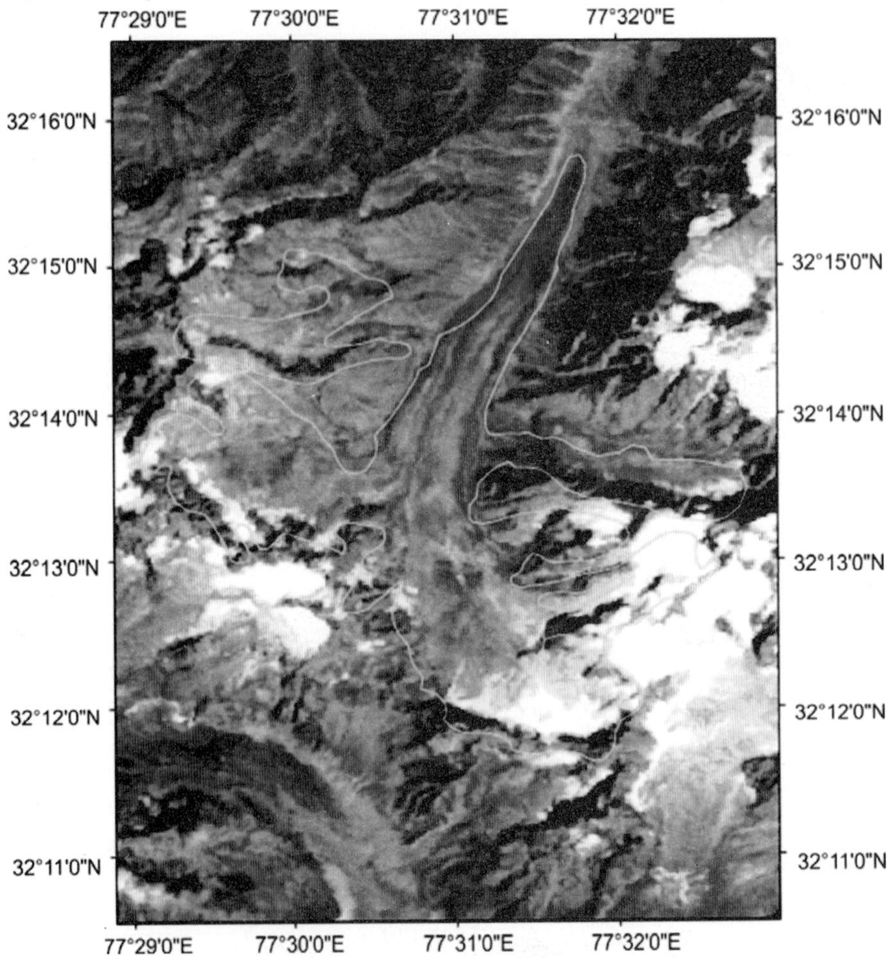

Fig. 1. The background is a LISS-III image of Chhota Shigri Glacier. The map coordinates are in the Geographic lat/long, WGS84 reference system.

Since the Himalaya presents marked seasonal temperature changes, glaciers in this region are summer ablation glaciers. While glaciers on the Tibetan plateau and in Nepal receive almost all precipitation from the summer monsoon and are thus summer accumulation glaciers. Chhota Shigri and other glaciers in Western Himalaya receive precipitation both from the summer monsoon and from mid-latitude westerlies in winter [Shekhar *et al.*, 2010]. Thayyen and Gergan (2010) distinguish between "Alpine catchments", "Himalayan catchments", and a cold-arid regime in Northwestern Himalaya. The Chandra basin, where Chhota Shigri is located, is an "Alpine catchment" dominated by winter precipitation, while monsoon precipitation becomes predominant in the "Himalayan catchments" located further south-east.

Table 1. Gives a list of geographical characteristics of Chhota Shigri Glacier.

General features	
State	Himachal Pradesh
Mountain range	Western Himalaya, Pir Panjal range
District	Lahaul and Spiti
Drainage system	Chandra River - Indus river (Chenab branch)
Climate	Monsoon-arid transition zone

Energy balance modelling

In very cold environments, sublimation replaces melt as the primary mechanism of surface ablation. To understand the causes of ablation by melting and sublimation, we need to examine the ways in which the glacier surface gains and losses heat and be able to measure or calculate the gains and losses under different weather conditions. The surface energy balance is an essential balance element of the climate in any region of the world. Small changes in the surface energy balance can lead to dramatic changes in the snow and ice cover itself [Luers and Bareiss, 2010].

Such study would be of practical value to hydroelectric companies concerned with predicting short term variations in melt water runoff in response to change in meteorological conditions. It can be used to study the relative importance of the energy balance components to surface melt changes spatially across a glacier and temporally throughout individual melt seasons in response to changing climate and surface conditions. Snowmelt runoff estimates are needed for forecasting seasonal water yields, river regulation and storage works, determination of design floods, planning flood control programs, etc [Arnold *et al.*, 1996].

The energy balance or the heat budget of a snowpack governs the production of melt water. This method involves an accounting of incoming energy, outgoing energy, and the change in energy storage for a snowpack for a given period of time and is controlled by meteorological conditions and physical properties of the glacier surface.

The energy balance of the glacier surface is described by the sum of the radiative (incoming and reflected solar radiation, incoming and outgoing long wave radiation) components and turbulent heat fluxes. Assuming no horizontal transfers of heat on the ground, the net energy flux into the surface can be expressed as:

$$Q = R + H + LE + G + P \qquad (1)$$

where Q is the energy available for melting of the snowpack; R is the net radiations; H is the turbulent sensible heat; LE is the turbulent latent heat of evaporation, condensation or sublimation, G is the conductive energy flux or subsurface energy flux in the snow/ice and P is the heat supplied by precipitation. Since precipitation is always snow in the vicinity of the equilibrium line and since snowfall intensities are usually weak, P remains insignificant and negligible as compared to the other terms. In equation (1) Q can also be expressed as:

$$Q = \Delta Q_m + \Delta Q_s$$

Here ΔQ_m is the latent heat storage change due to melting or freezing and ΔQ_s is the convergence or divergence of sensible heat fluxes within the snow pack volume [Wagon *et al.*, 1999]. Now the different components of energy balance (net radiations, turbulent fluxes and ground heat flux) in equation (1) can be elaborated as follows:

Net radiations

Solar radiation is the largest energy source and is able to change large quantities of liquid water into water vapor. Potential amount of radiation that can reach the evaporating surface is determined by its location and time of the year. Due to differences in the position of the sun, the potential radiation differs at various latitudes and in different seasons. The actual solar radiation reaching the evaporating surface depends on the turbidity of the atmosphere and the presence of clouds, which reflect and absorb major parts of the radiation. The net radiation is the balance of the incident and reflected short-wave radiation and the incident and emitted long-wave. It can be expressed as-

$$R = (1 - \alpha)\, G + (L_i - L_o)$$

According to Stefan's Boltzmann law above equation can be expressed as [Sellers *et al.*, 1997]-

$$R = (1 - \alpha)G + \varepsilon a \sigma Ta4 - \varepsilon s \sigma Ts4 \qquad (2)$$

where, α is albedo, G is global radiations, ε_a and ε_s are air and surface emissivities respectively, σ is Stefan's constant and its value is $5.67*10^{-8}$ W/m_2K_4 and T_a, T_s are air and surface temperatures respectively [Singh and Singh (2001). Snow and Glacier Hydrology]. All these parameters are detailed below:

Albedo

The conversion formula for the total shortwave broadband albedo for sensor Moderate Resolution Imaging Spectroradiometer (MODIS) is given below:

$$\alpha_{MODIS} = 0.160\alpha_1 + 0.291\alpha_2 + 0.243\alpha_3 + 0.116\alpha_4 + 0.112\alpha_5 + 0.081\alpha_7 - 0.0015$$

where α_1, α_2, α_3, α_4, α_5 and α_7 are spectral narrowband albedos in I, II, III, IV, V and VII bands respectively [Liang, 2000].

Global radiations

For clear sky G can be calculated using the following equation [Samani *et al.* 2007]:

$$GG = as + bs\ nnNN\ RRaa \quad (3)$$

where n is actual duration of sunshine (hour), N is maximum possible duration of sunshine or daylight hours (hour), n/N is relative sunshine duration, R_a is extraterrestrial radiation (M J m^{-2} day^{-1}), a_s is regression constant, expressing the fraction of extraterrestrial radiations reaching the earth on overcast days (n=0), $a_s + b_s$ fraction of extraterrestrial radiation reaching the earth on clear days (n=N). Commonly used values for a_s and b_s are 0.25 and 0.50 respectively. Equation number (3) for the clear sky radiation reaching the earth on clear sky days (n=N) can be written as :

$$GG = (as + bs)\ RRaa \quad (4)$$

Now the extraterrestrial radiation, R_a, for each day of the year and for different latitudes can be estimated from the solar constant, the solar declination angle and the time of the year by the following formulation:

$$RRaa = 24 * 60\ \pi\ GSC * dd_{rr}\ (\omega\omega ssSSSSSS\phi\phi\ ssssss\delta\delta + cccccc\phi\phi\ cccccc\delta\delta\ ssssss\omega\omega ss) \quad (5)$$

where Gsc is solar constant and given as 0.0820 M J m^{-2} min^{-1}, d_r is inverse relative earth-sun distance and is calculated as:

$$dr = 1 + 0.033\ cccccc\ (2\pi JJ|365) \quad (6)$$

where J is Julian day of the year, ω_s is sunset hour angle given by:

$$\omega\omega ss = cccccc - 1\ (-\ tttttt\phi\phi *\ tttttt\delta\delta) \text{ in radians} \quad (7)$$

Φ is latitude in radians, δ is solar declination (rad), which is calculated as:

$$\delta\delta = 0.409 * ssssss2\pi JJ/365 - 1.39 \quad (8)$$

by substituting values of parameters from equations (6), (7) and (8) in equation (5), equation (4) can be expressed as:

$$G = 0.75 * Ra \text{ in M J m}^{-2} \text{ day}^{-1} \quad (9)$$

Incoming long wave radiations

They are emitted by the atmosphere, primarily by water vapor, CO_2 and ozone. Wavelength range is 4-120 μm. During higher temperature and more cloudy conditions, long wave radiations are high. It is calculated as [Hock, 2010] :

$$Li = \varepsilon a \sigma Ta4 \quad (10)$$

Here ε_a is calculated using following formulation:

$$\varepsilon\varepsilon_{aa} = 0.7+5.95*10-5* ee_{aa} \exp(1500TT_{aa}) \tag{11}$$

where $\varepsilon\varepsilon_{aa}$ emissivity of air and T_a, e_a are surface air temperature in K and vapor pressure in kPa at that temperature at about 1.0-2.0 m above the surface respectively. T_a is a critical variable to describe the energy and water cycle of the Earth-atmosphere system and is a key input element for energy balance modelling. Based on the linear regression relationship between surface air temperature obtained from AWS and MODIS land surface temperature, surface air temperatures are estimated from 1 km MODIS land surface temperature. Vapor pressure at a given air temperature is calculated as:

$$ee_{aa}= RR_h * ee_{ss}100 \tag{12}$$

where R_h is relative humidity and is obtained through AWS data and e_s is saturated vapor pressure and it is related with T_a through relation given below:

$$ee_{ss} = 6.11\exp(17.27TT_{aa} \mid 237.3 + TT_{aa}) \tag{13}$$

and e_s is in kPa and σ is Stefan Boltzmann constant and its value is $5.67 * 10^{-8}$ W/m^2

Outgoing long wave radiations

Long wave radiations are emitted by earth's surface. It is a function of temperature of the ice/snow surface and it can't exceed 316W/m^2 because ice/snow can't be warmer than zero degrees. It is calculated as:

$$Lo = \varepsilon s\sigma Ts_4 \tag{14}$$

where ε_s is surface emissivity taken from literature as 0.965 (average value) and T_s is land surface temperature obtained as (MOD11A1) data product from Terra-MODIS.

Sensible and latent heat fluxes

Bulk aerodynamic method was used to calculate turbulent fluxes of sensible and latent heat. This method is often used for practical purposes as it approximates turbulent heat fluxes from one level of measurement. In this method, a constant gradient is assumed between the level of measurement and surface and thus surface values have to be evaluated. In this approach, the stability of the surface layer is assessed by calculating the bulk Richardson number, R_b, which relates the relative effects of buoyancy to mechanical forces [Reid and Brock, 2010] -

$$RR_{bb} = g\,(TT_{aa} - TT_{ss})\,(zz_{aa} - zz0_{\mathrm{mm}})\,TT_{mmuu2} \tag{15}$$

where T_a and u are the values of air temperature (in K) and horizontal wind speed (in m/s) respectively at the level of measurement z. g is the acceleration due to gravity ($g = 9.8$ m s^{-2}). T_s is the surface temperature (in K). T_m is the mean absolute air temperature between the surface and the measurement level z (in K). z_{0m} is the surface roughness length for momentum (in m). The roughness length is defined as the height above a surface at which the extrapolated horizontal wind speed profile reaches zero.

Assuming that local gradients of mean horizontal wind speed u, mean air temperature T_a and mean vapor pressure are equal to the finite differences between the measurement level and the surface, it is possible to give analytical expressions for the turbulent fluxes. [Oke, 1987]:

$$HH=\rho C_P kk_2(TT_{aa}-TT_{ss})\ llll_{zzaazz0mm}\ lll_{zzaazz0tt}\ (\phi\phi mm\ \phi\phi h)-1 \quad (16)$$

$$LLL=\rho L_v k_2(qq_{aa}-qq_{ss})\ llll_{zzaazz0mm}\ lll_{zzaazz0tt}\ (\phi\phi mm\ \phi\phi vv)-1 \quad (17)$$

where q_a and q_s are specific humidities at the level of measurement z and surface, respectively. ρ is the air density; C_P is the specific heat capacity for air at constant pressure; k is the von Karman's constant (k = 0.4) and L_v is the latent heat of vaporization (L_v = 2.476 * 10_6 J/kg). The surface roughness lengths for heat z_{0t} and humidity z_{0q} were considered equal to z_{0m}. The nondimensional stability functions for momentum (Φ_m), heat (Φ_h) and moisture (Φ_v) are expressed as functions of R_b [Brutsaert, 1982; Oke, 1987]:

For a stable surface layer, $R_b \geq 0$

$$(\Phi_m\Phi_h)^{-1} = (\Phi_m\Phi_v)^{-1} = (1-5R_b)_2 \quad (18)$$

For an unstable surface layer, $R_b < 0$:

$$(\Phi_m\Phi_h)^{-1} = (\Phi_m\Phi_v)^{-1} = (1-16R_b)_{0.75} \quad (19)$$

In equation (16) and (17), wind speed at height z is obtained from AWS data. Land surface temperature is obtained from Terra-MODIS. For the month of November, 2011 roughness length for momentum transfer (z_{0m}) is taken from literature, recorded over mid and low latitude glaciers for fresh snow surfaces as 0.2 mm and for September, 2011 it is measured using surface profilometer during field work and then applying microtopographic method. Several workers have sought to overcome the z_{0m} measurement problem through microtopographic measurements of the glacier surface. The empirical relationship [Lettau, 1969] is used for calculating z_0 as:

$$z_0 = 0.5h^*(s/S) \quad (20)$$

In which h* is the average vertical extent, or effective obstacle height, of the roughness elements; s is the silhouette area (area of upwind face of an average element) and S is the unit ground area occupied by each element. The only measurements required are of the variation of surface elevation made at regular intervals relative to a horizontal reference in a plane perpendicular to the prevailing wind; h* is calculated as twice the standard deviation of the elevations (σ), with the mean elevation set to zero. The number of continuous groups of positive height deviations above the mean elevation defines the frequency, f; of roughness elements and the width of a typical element is defined as the length of the traverse, X, divided by 2f [Brock *et al.*, 2006]. Equation number (20) is solved by substituting s = h*X/2f and S = $(X/f)_2$. So the equation (20) becomes:

$$z_0 = 0.5h^*{}_2f/2X \quad (21)$$

according to prevoius explanation equation (21) becomes-

$$z_0 = \sigma_2 f/X \quad (22)$$

Ground heat flux

The ground heat flux or the conductive heat transfer within the snowpack or the ice tends to be small when compared to radiative or turbulent fluxes [Marks and Dozier, 1992]. The thermal energy stored by the ground during the summer period, when there is no snow cover over the ground, is released during winter and spring which contributes to the melting of the overlying snowpack. The temperature of the ground surface is reduced due to existing snowpack in comparison to the lower part of the ground, which results in a temperature gradient. Thus the heat is supplied by ground in the upward direction. In case the temperature of snow just above the ground is below 0°C, the ground heat flux raises the temperature and makes it ready for melting. The vertical flux of heat at the snow surface is described as :

$$GG = -KK\,\partial\partial TT\partial\partial zz \qquad (23)$$

where K is the thermal conductivity of snow/ice (in W m^{-1} K^{-1}), T is the snow/ice temperature, and z is the depth [Wagon *et al.* 1999].

Surface energy balance of snowpack

The internal energy change of the snow pack (ΔQm + ΔQs) is calculated using equation (1). If the left hand side of the equation (1) is positive, energy is available for the snow/ice. It will be used first to increase the snow/ice temperature until it reaches 0°C (ΔQs >0 and ΔQm = 0), when the surface temperature reaches 0°C, melting occurs (ΔQm > 0). If the LHS is negative then the reverse situation is observed, it indicates that the meltwater of the snow pack refreezes (ΔQm < 0) and then snow/ice temperature decreases (ΔQs < 0). ΔQs is the rate of gain/loss of heat in a vertical column extending from the surface to the depth of snowpack. Therefore, considering daily means, ΔQs usually remains zero because the gain of heat during the day is compensated by the loss of heat at night. Therefore, looking at daily means, the change of the internal energy of snowpack is reduced to the latent heat storage change. Now Q_m is converted to mass units using the Latent heat of fusion L_f ($L_f = 3.34 * 10^5$ J/kg).

$$\Delta Q_m = L_f * M$$

where M is the melt rate in mass per unit area per unit time (kg m^{-2} s^{-1} or mm/s). Similarly the sublimation rate can be calculated using following relation:

$$LE = L_s * S$$

where LE is the Latent Heat Flux calculated by equation (17) and L_s is the Latent heat of sublimation L_s ($L_s = 28.3 * 10^5$ J/kg)

So the ablation (A), can be calculated as:

$$A = M + S$$

Data used

The input parameters for the energy balance modelling have been collected from the following sources:

Satellite data

Surface reflectance and land surface temperature data products at 500m and 1km resolution respectively of MODIS, onboard TERRA have been used in the study. It is an Earth Observing satellite, launched on 20 December, 1999 with orbital height 705 km and swath of 2330 km (cross track by 10 km (along track at nadir). It is having 36 spectral bands with spatial resolution as: 250m (band 1 – 2), 500m (band 3 – 7), 1000m (band 8 – 36) and quantization is 12 bits.

Meteorological data

Meteorological parameters such as wind speed, air temperature and relative humidity are taken from the AWS (Automatic Weather Station), shown in Fig. 2, installed at Chhota Shigri glacier at 3851.46m above sea level.

Fig. 2. AWS installed at Chhota Shigri

Field data

The surface roughness parameter used to calculate turbulent heat fluxes is obtained by doing field measurements using surface profilometer. The surface profilometer is a mechanical system designed to measure roughness profile by microtopographic method. As shown in figure-3 such type of 6 aluminum frames were designed with each having width of 50 cm and length of 90 cm. The spacing between two consecutive rods is 2 cm, diameter of the rod is 0.5 cm and length of the rod is 50 cm. On the tip of rods bolts were welded so that the rods do not fall while lifting the instrument. And a total of 24 rods were used with each frame, so that it provides snow/ice roughness profile measurements over 3 m length. To support the instruments, angle of same size were used at the base. The roughness length is an important control on the rate of turbulent heat transfer between a glacier surface and the air above it. The lack of systematic monitoring of z_0 variations on glaciers stems partly from the difficulty of recording z_0 at a large number of different sites, since techniques based on measurement of airflow in the surface atmospheric boundary layer (SABL) require long periods of monitoring to generate a single z_0 value and thus the microtopographic method has a distinct advantage over other roughness length measurement methods. It enables repeated measurements to be made at many points across a glacier surface. As mentioned earlier in microtopographic method, the empirical relati-onship between the surface roughness element geometry and z_0 given by Lettau as:-

$$z_0 = 0.5h^*(s/S)$$

As the surface roughness length for momentum transfer (z_{0m}), the surface roughness length for heat (z_{0t}) and humidity (z_{0q}) were considered equal so it can be expressed as:

$$z_{0m} = z_{0t} = z_{0q} = z_0$$

Figure 3 depicts the manually measured surface microtopography in the ablation zone of Chhota Shigri glacier, using 3m frame (joining rest of 5 frames in series to generate 3m profile at a horizontal interval of 2 cm). As described in the microtopographic method, the only measurements required are of the variation of surface elevation made at regular intervals. For this, 152 vertical distance were measured (Fig. 4) and then the mean elevation was set to zero. In this way the h* in equation (20) is calculated as twice the standard deviation of the elevations so by calculating the value of h*, s and S through roughness profile we put these values in equation (22) and for the month of September, 2011 we calculated surface roughness over the ablation zone of Chhota Shigri Glacier as:

$$z_0 = 4.484\ mm$$

As reported in literature that the roughness length for rough glacier ice surface recorded over mid and low latitude glaciers by eddy covariance method, microtopographic method and wind profile method, comes out to be of order of

3-15 mm range. So the above roughness length value is validated with the values reported in literature.

Fig. 3. Schematic figure of surface profilometer

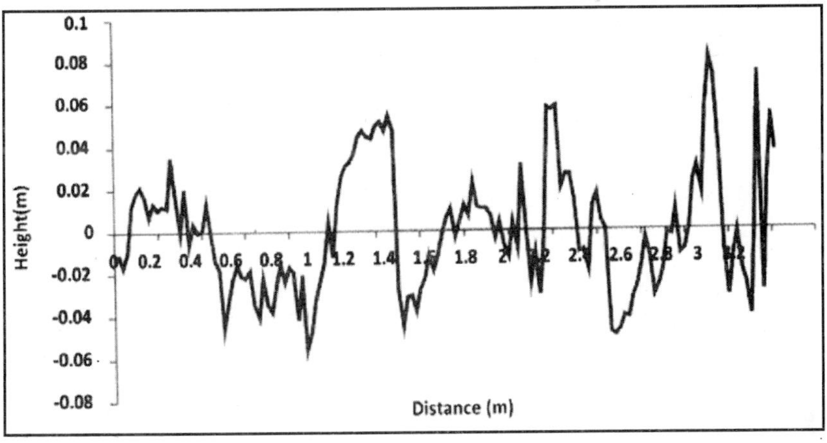

Fig. 4. Roughness profile for the ablation zone of Chhota Shigri glacier.

Results and discussions

Net radiations

The surface energy balance terms for the entire measuring period (November, 2011) are shown in figures in which a number of interesting features can be noted.

Figure 5 is showing the daily mean variation of net shortwave radiations throughout the month of november, 2011. The daily average net shortwave radiation comes out to be nearly 89.67 W/m^{-2}. Positive sign signifies that it is the source of energy at the glacier surface. The fluctuations in net shortwave radiations are due to variations in albedo values and solar insolation. Figure 6 is showing the daily mean variation of net longwave radiations throughout the month of November, 2011. The daily average net longwave radiation comes out to be nearly -42.34 W/m^{-2}. The fluctuations in net longwave radiations are due to variations in air and surface temperatures. It shows the negative values because of high radiative cooling of the surface and thus it represents heat sink. The net radiation is the difference between total upward and downward radiation fluxes on the surface or the sum of net shortwave and net longwave radiations. Figure 7 is showing the daily mean variation of net radiations throughout the month of November, 2011. The daily average of net radiation comes out to be nearly 50.045 W/m^{-2}. Positive sign signifies that it is the source of energy at the glacier surface.

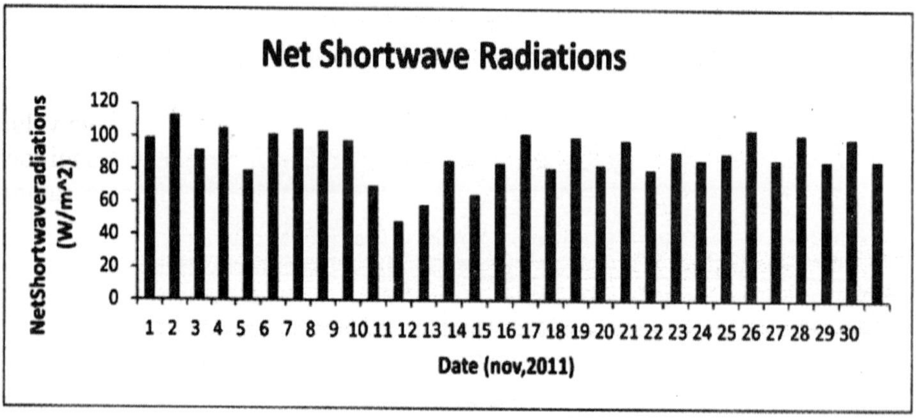

Fig. 5. Daily variation of net shortwave radiation during November, 2011

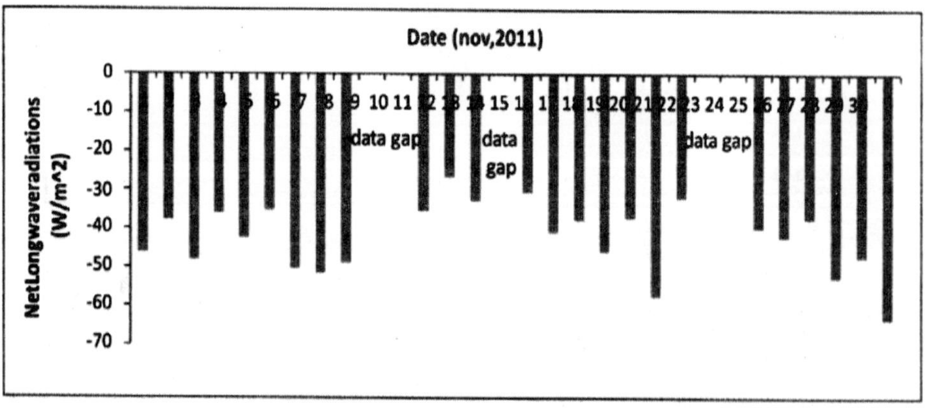

Fig. 6. Daily variation of net longwave radiation during November, 2011

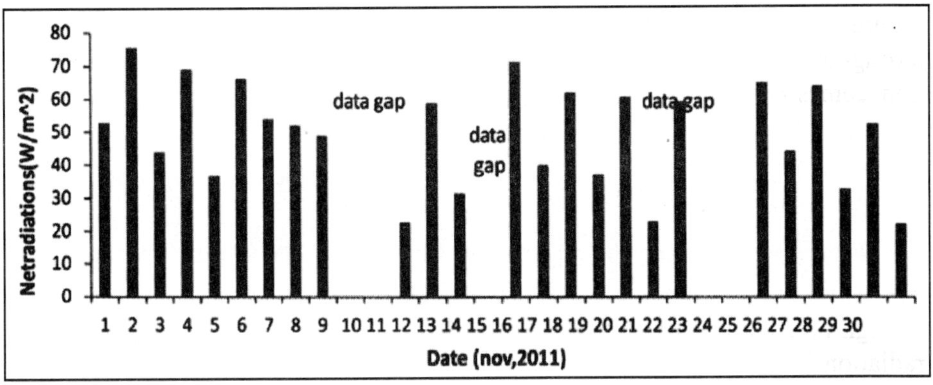

Fig. 7. Daily variation of net radiation during November, 2011

Sensible heat flux

The transport of heat and moisture in the surface boundary layer of the atmosphere is dominated by turbulent motions and that is why sensible and latent heat fluxes are called turbulent heat fluxes. Figure 8 is showing the daily mean variation of sensible heat flux throughout the month of November, 2011. The daily average of sensible heat flux comes out to be nearly 14.72 W/m^{-2}. Positive sign signifies that it is the source of energy at the glacier surface. Positive values also signify that the surface boundary layer is almost in stable conditions. Sometimes the very high fluxes result of high winds and warm temperatures. Sensible heat is non-negligible but remains small during the wet season. As reported in literature that unstable conditions are sometimes encountered in the early morning (around 0800 – 0900 LT) while snow/ice surface is heated by incident shortwave radiation and air temperature is still negative. In these cases rather typical of the dry season, sensible heat flux is negative. After a while, the temperature at the surface reaches its upper limit 0°C, air temperature keeps increasing to positive values and sensible heat flux is negative and then the lower atmosphere becomes stable with positive sensible heat flux [Wagon *et al.,* 1999].

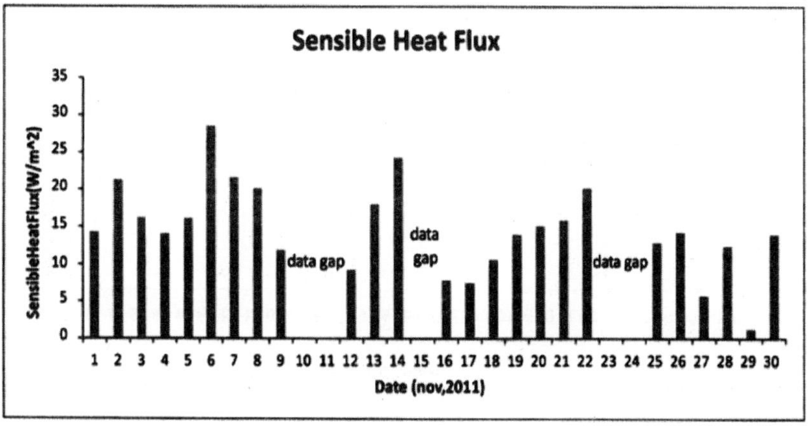

Fig. 8. Daily variation of sensible heat flux during November, 2011

Latent heat flux

The highly variable wind speed, air temperature and specific humidity on a time scale of few days cause strong fluctuations in latent heat fluxes. Indeed, in the morning, as soon as the Sun rises, the atmosphere is dry and latent is highly negative. When the wet air masses invade the glacier, the relative humidity arises rapidly and latent heat drops

Figure 9 is showing the daily mean variation of latent heat flux throughout the month of November, 2011. The daily average of latent heat flux comes out to be nearly -5.48 W/m^{-2} and negative sign represents heat sink. Negative values of latent heat flux also indicate that snow/ice surface loses mass by sublimation and also infers that the stratification of the lower atmosphere is moderately stable.

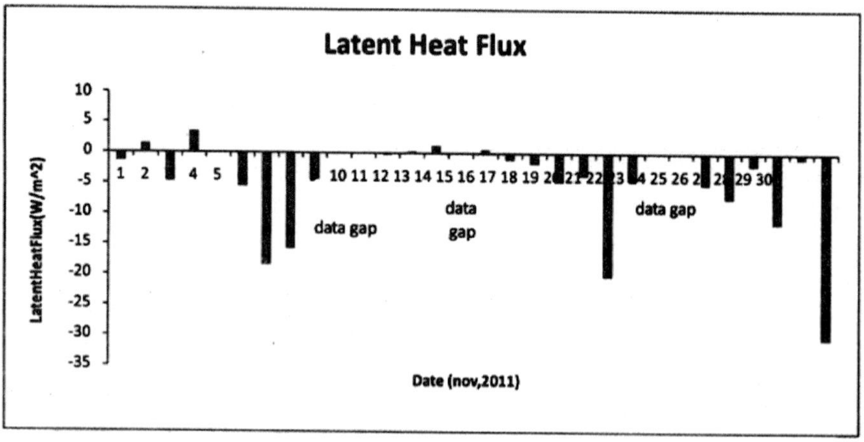

Fig. 9. Daily variation of latent heat flux during November, 2011

Figure 10 is showing the daily mean variation of melting throughout the month of November, 2011. The daily average melting comes out to be nearly 15.33 mm water equivalent. The fluctuations in the melting are because of the variation of the total energy received at the glacier surface. Higher be the energy received at the surface higher the melting takes place.

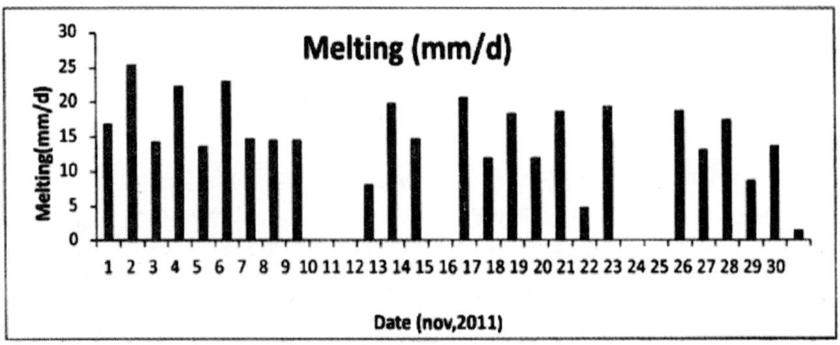

Fig. 10. Daily variation of melting during November, 2011

Figure 11 is showing the daily mean variation of sublimation throughout the month of November, 2011. The daily average sublimation comes out to be nearly 0.1675 mm water equivalent. The fluctuations in the sublimation are because of the variation of the latent heat flux at the glacier surface. As most of the times the latent heat flux is negative which corresponds to sublimation. Positive values of latent heat flux are corresponding to condensation on the glacier surface.

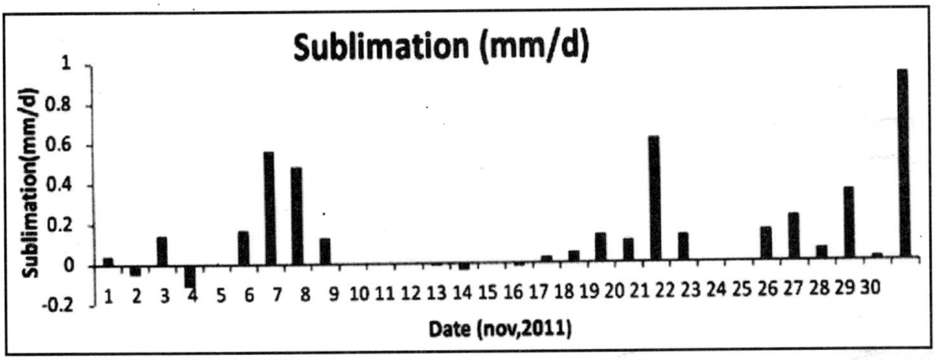

Fig. 11. Daily variation of sublimation during November, 2011

Figure 12 is showing the daily mean variation of ablation throughout the month of November, 2011. The daily average ablation comes out to be nearly 15.50 mm water equivalent. The fluctuations in the ablation are because of the variation of the melting and sublimation at the glacier surface, as ablation is sum of melting and sublimation.

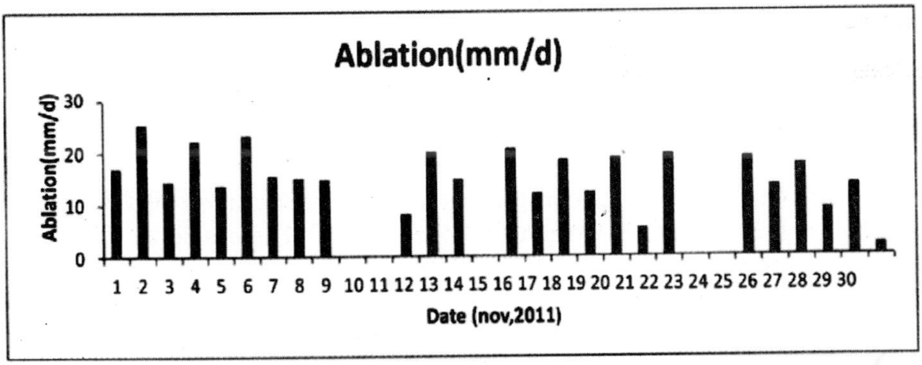

Fig. 12. Daily variation of ablation during November, 2011

Summary and conclusions

Physically based models of the glacier surface energy balance (SEBM) account for the heat inputs and losses to a glacier surface and are detailed enough to capture most of the physical processes that govern surface melt. Surface energy balance calculations have been made over glaciers at the point scale [Munro, 1990; Marks and Dozier, 1992; Greuell and Konzelmann, 1994; Brock and Arnold, 2000], along centerlines [Oerlemans and Hoogendoorn, 1989; Oerlemans,

1992] and in spatially distributed form [Escher-Vetter, 1985; Arnold *et al.*, 1996; Kolk and Oerlemans, 2002; Hock and Holmgren, 2005]. These models differ from the commonly used empirical ablation models such as the Positive Degree Day (PDD) model of Braithwaite [1981], the temperature/solar radiation index model developed by Hock [1999] and the enhanced temperature index model developed by Pellicciotti *et al.* [2005], which use empirical relationships between several (i.e., temperature, solar, precipitation) meteorological variables and surface melt. Because SEBMs can be formulated to be driven by a relatively small set of meteorological data, they are readily applied anywhere forcing data are measured or obtainable. As this type of surface energy balance modelling has not been done in India yet, so this study is an attempt to assess the surface energy balance components over Indian glaciers.

Towards understanding Chhota Shigri Glacier's response to climate forcing, this study revealed the basic characteristics of the energy balance on a glacier surface. The different energy components (net radiations, sensible heat flux and latent heat flux) are calculated at the glacier surface using the satellite data (surface reflectance, land surface temperature), meteorological data (relative humidity, wind speed and air temperature) and field data (surface roughness using surface profilometer), through which melting and sublimation is calculated and then melting and sublimation contribute to ablation. Energy balance is dominated by radiative exchange which is governed by the variation in net shortwave radiation. Turbulent latent heat flux is negative most of the time during the measuring period which shows continuous sublimation and positive values correspond to condensation on the glacier surface. Turbulent sensible heat flux is little bit prominent in comparison to latent heat flux, which indicates the marginal role of local air tempertature by itself on ablation. During the period under consideration (November, 2011) daily average values of net shortwave radiations (89.67 W/m^2) and the sensible turbulent heat flux (14.72 W/m^2) are energy sources at the glacier surface, whereas the net longwave radiations (-42.34 W/m^2) and the latent turbulent heat flux (-5.489 W/m^2) represent heat sinks. Sum of net shortwave and net longwave is the net radiation (50.045 W/m^2). Daily average value for the month of November, 2011, the amount of energy available for the melting of glaciers is 59.28 W/m^2; mean melting rate from the glacier surface is 15.33 mm water equivalent/d; mean sublimation rate over the glacier surface is 0.1675 mm water equivalent/d and thus mean ablation rate is computed as 15.50 mm water equivalent/d. In this way the result of this study facilitate a better understanding of the glacier's dynamics and the response of glaciers to change in temperature.

Acknowledgements

Authors are thankful to Shri. A.S. Kiran Kumar, Director, SAC for his interest in this study and research facilities.

References

Arnold, N.S., Willis, I.C., Sharp, M.J., Richards, K.S. and Lawson, W.J., A distributed surface energy balance model for a small valley glacier, *Journal of Glaciology*, 42 (1996) 77-89.

Anslow, F.S., Hostetler, S., Bidlake, W.R. and Clark, P.U., Analysis of meteorological data and the surface energy balance of McCall Glacier, Alaska, USA, *Journal of Glaciology*, 51 (2005) 451-461.

Braithwaite, R.J., On glacier energy balance, ablation, and air temperature, *Journal of Glaciology*, 27 (1981) 381-391.

Brock, B.W. and Arnold N.S., A spreadsheet- based point surface energy balance model for glacier and snow melt studies, *Earth Surf. Processes Landforms*, 25 (2000) 649-658

Brock, B.W., Willis, I.C. and Sharp, M.J., Measurement and parameterization of roughness length variations at Haut Glacier, *Journal of Glaciology*, 52 (2006) 1-17.

Brutsaert, B., Evaporation in the Atmosphere, Theory, History and Applications, (Kluwer Acad., Norwell, Mass) 1982, pp.299.

Escher-Vetter, H., Energy balance calculations for the ablation period 1982 at Vernagtferner, Oetzal Alps, *Ann. Glaciol*, 6 (1985) 158-160.

Greuell, W. and Konzelmann, T., Numerical modeling of the energy balance and the englacial temperature of the Greenland Icesheet: Calculations for the ETH-Camp location (West Greenland, 1155 m a.s.l.), *Global Planet. Change*, 9 (1994) 91-114.

Hock, R., A distributed temperature-index ice and snowmelt model including potential direct solar radiation, *Journal of Glaciology*, 45 (1999) 101-111.

Hock, R. and Holmgren, B., A distributed surface energy balance model for complex topography and its application to Storglaciaren, *Journal of Glaciology*, 51 (2005) 25-36.

Hock, R., Glacier meteorology energy balance, *Summer school in glaciology*, (2010) 1-10.

Kolk, E.J. and Oerlemans, J., Model study of the spatial distribution of the energy balance of Morteratschgletscher, Switzerland, *Journal of Glaciology*, 48 (2002) 505-518. .

Luers, J. and Bareiss, J., The effect of misleading surface temperature estimations on the sensible heat fluxes at a high Arctic site, *Atmos. Chem. Phys.*, 10 (2010) 157-1.

Lettau, H., Note on Aerodynamic Roughness- Parameter Estimation on the basis of roughness-element description, *Journal of Applied Meteorology*, 8 (1969) 828-832.

Liang, S., Narrowband to broadband conversions of land surface albedo I algorithms, *Remote Sensing of Environment*, 76 (2000) 213-238.

MacDouall, A.H. and Flowers, G.E., Spatial and temporal transferability of a distributed energy balance, *Journal of Climate*, 24 (2010) 1480-1498.

Munro, D.S., Comparison of melt energy computations and ablatometer measurements on melting ice and snow, *Arct. Alp. Res.*, 22 (1990) 153-162.

Marks, D. and Dozier, J., Climate and energy exchange at the snow surface in the Alpine Region of the Sierra Nevada: 2. Snow cover energy balance, *Water. Resour. Res.*, 28 (1992) 3043-3054.

Oke, T.R., *Boundary Layer Climates*, (Routledge, New York) 1987, pp.435.

Oerlemans, J. and Hoogendoorn, N.C., Mass balance gradients and climate change, *Journal of Glaciology*, 35 (1989) 399-405.

Oerlemans, J., Climate sensivity of glaciers in southern Norway: Application of an energy balance model to Nigardsbreen, Hellstugubreen and Alfotbreen, *Journal of Glaciology*, 38 (1992) 223-232.

Pellicciotti, F., Brock B., Strasser U., Burlando, P., Funk, M. and Corripio, J., An enhanced temperature-index glacier melt model including the shortwave radiation balance: Development and testing for Haut Glacier d'Arolla, Switzerland, *Journal of Glaciology*, 51 (2005) 573-587.

Reid, T.D. and Brock, B.W., An energy balance model for debris covered glaciers including heat conduction through the debris layer, *Journal of Glaciology*, 56 (2010) 903-916.

Shekhar, M., Chand, H., Kumar, S., Srinivasan, K. and Ganju, A., Climate change studies in the Western Himalaya, *Ann. Glaciology*, 51 (2010) 105-112.

Sellers, P.J., Dickinson, R.E., Randall, D.A., Betts, A.K., Hall, F.G., Berry J.A., Collatz, G.J., Denning, A.S., Mooney, H.A., Nobre, C.A., Sato, N., Field, C.B. and Sellers, A.H., Modeling the exchanges of energy, water, and carbon between continents and the atmosphere, *Journal of Glaciology*, 51 (1997) 25-36.

Samani, Z., Bawazir, A.S., Bleiweiss, M., Skaggs, R. and Tran, V.D., Estimation daily net radiations over vegetation canopy through remote sensing and climatic data, *Journal of Irrigation and drainage engineering*, 133 (2007) 291-297.

Singh P. & Singh V.P., *Snow and Glacier Hydrology*, (Kluwer Academic, The Netherlands) 2001, pp.743.

Thayyen, R. J. and Gergan, J.T., Role of glaciers in watershed hydrology: A preliminary study of a Himalayan catchment, *The Cryosphere*, 4 (2010) 115-128.

Wagon, P., Ribstein, P., R.E., Francou, B. and Pouyaud, B., Annual cycle of energy balance of Zongo Glacier, Cordillera Real, Bolivia, *Journal of Geophysical Research*, 104 (1999) 3907-3923.

Natural hazards due to climate change in Himalaya and multicriteria decision analysis to identify vulnerable zone

[1]Pankaj Gupta, [2]J. Sundaresan, & [3]K.M. Santosh*

[1]Geotechnical Engineering Division, CSIR- Central Road Research Institute, Delhi - Mathura Road, New Delhi 110 025
[2]Climate Change Informatics, CSIR-NISCAIR, New Delhi-110 012, India
School of Marine Sciences, CUSAT, Kochi 682 016, India
*[E.mail: sundaresanj@hotmail.com, panx22@gmail.com]
*For correspondence

Introduction

Glaciation in Himalayan mountain chain is developed about two million years ago since the late ice age in Pleistocene Period (Kuhle, 2011). Himalayan region has the largest ice cover in terms of numbers of glaciers (Jianchu & Grumbine 2009). The present inter-glaciers warm period is a period of glaciers retreat (Lehmkuhl & Owen 2005). Glacier located in the Central Himalaya has recorded an annual retreat of 8-10 m/y (Raina, 2010). This may be due to complex regional and local phenomena along with climate change [World Glaciers Monitoring Service (WGMS)]. Glaciers retreat has been accelerating the gain since 1980's beyond the range of pre-industrial variability (WGMS). However, the studies related to abnormal increase in glacier retreat at Himalaya and subject like whether glaciers will respond to immediate change in global warming due to climate change are in an infant stage.

Glacier melting is an imminent threat due to climate change in Himalayan region. This will result in increase in water level of glacier lakes, water recharge in watershed region of the glaciers fed rivers and sloppy fracture zones of the aquifers rocks. Glacial retreat will encourage avalanches, landslides in this region (Fig. 1). It may have catastrophic effects on the glaciers fed rivers in general and watershed area of the river in particular. Pindari, Kafani, Mrigthuni and Mangtoli Glaciers are the four glaciers which fed the Pindari River and tributary of Alaknanda River. Alaknanda is also one of the main tributary of Ganga River. The watershed of Pinder River is in the buffer zone of the NDBR. Studies to minimise the impact of climate change in Himalaya is not so conspicuous. The primary task for any adaptation and mitigation programme is to identify the vulnerable zone of the region. There are attempts to delineate global climate change and its impacts on Himalayan glaciers (Vince, 2010, Whipple, 2009, Wobus, Heimsath, *et al.*, 2005, Burbank, 2005, Hewitt, 1988, Brozović, Burbank and Meigs, 1997, Korup *et al.*, 2010, Huntington, Blythe and Hodges, 2006, Kothawale and Kumar, 2005, Pradhan, Singh, Buchroithner, 2006, Kumar and Mehrotra, 2011, Jiawen *et al.*, 2004, Kulkarni *et al.*, 2007, Bali *et al.*, 2011, Kireet *et al.*, 2008). Studies to establish the signatures of hazardous phenomena through vulnerable region are rare. In the present study drainage characteristics were examined through Remote Sensing and GIS technique to assess the hydrological characteristics and the vulnerable zones of Indian Himalayan region in the perspectives of climate change.

Himalayan region is highly prone to many hazards viz. earthquake due to its neo tectonic activities, cloud burst, avalanches. These hazards are capable enough to bring down any catastrophic landslide in the area, made up of fragile litholgy. These catastrophic landslides are also capable to block the river stream for couple of hours to even couple of months. These temporary natural dams have very low stability in general. The enormous water impounding behind these dams may topple without any prior warning. Breaching of such temporary dams may flood the downstream side. Water from such toppled dams is enormous with soaring speed due to steep gradient, capable to carry huge rock load, further increases its destructive erosion capacity.

Present study consist the Multicriteria Decision analysis (Strahler, 1964) to understand the morphometry of catchment area and to identify the vulnerable zones of the specific regions. Thematic maps viz. digital elevation model, relative relief map, slope map, aspects map, drainage density, stream frequency, fracture density were developed. Weight and score for each thematic layer and their class has been estimated and applied. Morphometric parameters were estimated and utilised to understand the hydrological behaviour and topography of the study area.

Material and methods

The study area consisting Pindari River Basin (PRB) has unique combination of mountain ecosystem including traditional agro ecosystem, mixed temperate, sub alpine and glaciers. Study area *(Nu = 5, A = 557.63 km^2, L = 34.13 km*; N. *Lat 30°03" 23"- 30°19' 04" and 79°45" 59"- 80°04'13")* (Fig. 1) falls into a tectonically active zone of Indian Himalayan Region, with a complex geotectonic set up confined by a number of thrusts and faults (Pradhan *et al.,* 2008). It consist the Inner Himalaya (Lesser Himalaya), the Himadri (Grate Himalaya) and The Tethys Himalaya. Rocks belonging to the Vaikrita and Almora Groups are mostly covered by the glaciers. Various rock formations belonging to the corresponding Groups are exposed in the study area. Main Central Thrust (MCT) is passing through the area separating the rocks of Vaikrita Group from the rocks of Almora Group. Similarly, rocks of the Jaunsar Group are thrusted under the rocks belonging to Almora Group by Munsiari Thrust in this region. Rocks of Tejam Group are also separated by a thrust from the Dhamta Group (Valdiya, 2010). General Geological Succession of the area is shown in Table 1.

Table 1. Geological Succession of the area (Valdiya, 1980)

Group	Formations	Main Rock Type	Relative Age
Jaunsar	Berinag	Quartzite with amygdaloidal & vesicular basalts	Youngest ↑
Tejam	Chandpur	Phyllites interbedded with metasiltstone and very fine grained wackes	
	Mandhali (Sor+Thadeddar)	Slate, Phyllite, Limestone and dolomite etc.	
Damtha	Rautgara	Limestone, Dolomite, calc slate and dolerite sill and dykes	
Almora	Saryu-Gumalikhet	Chlorite-biotite schist and biotite quartzite and augen gneisses	
Vaikrita	Joshimath	High Grade metamorphic rocks (Kyanite-sillimanite-garnet bearing gneiss and schist, garnet-quartx-mica granulite and calc-silicate rocks)	Oldest

Study area falls in a typical Himalayan region. Various rock types, High Grade metamorphic rocks (Kyanite-sillimanite-garnet bearing gneiss and schist, garnet-quartz-mica granulite and calc-silicate rocks) to low metamorphic rocks (Phyllites) are exposed in the area. These rock units are thrusted and separated by the number of major thrusts viz. MCT, Munsiari etc. Rocks exposed in the area are folded, faulted, sheared and highly jointed because of the tectonic activities. These rocks are further weak and vulnerable for erosion and weathering due to the high rainfall, steep slope.

Multicriteria analysis: The procedure ascribed in Mather & Koch, 2011 is utilised in the study. Thematic layers (slope, drainage density, fracture density, relative relief, and drainage frequency) ranked as equal weights and multiplied by their corresponding scores. These products were summed up to obtain the vulnerable zone. It is overlaid with geological and aspect map to assess the impact. This was done by utilizing the arithmetic overlay (weighted sum) built into ArcGIS model builder. Thus each alternative can be evaluated using the following equation.

$$Uig = \sum \lambda g \, \mu g \, (X)$$

where,

λg = is a positive scaling constant or the weight given to the gth individual

μg= is a utility function for the gth person over decision alternatives x, scaled from 0 -1

Weights and scores for thematic layers and their classes were estimated based on the method followed by Mather & Koch, 2011. We had modified the parameters based on the vulnerability of the study area. In the present study it is assumed that medium drainage density, high fracture density, high slope and high relative relief with SW-SE aspect are more vulnerable.

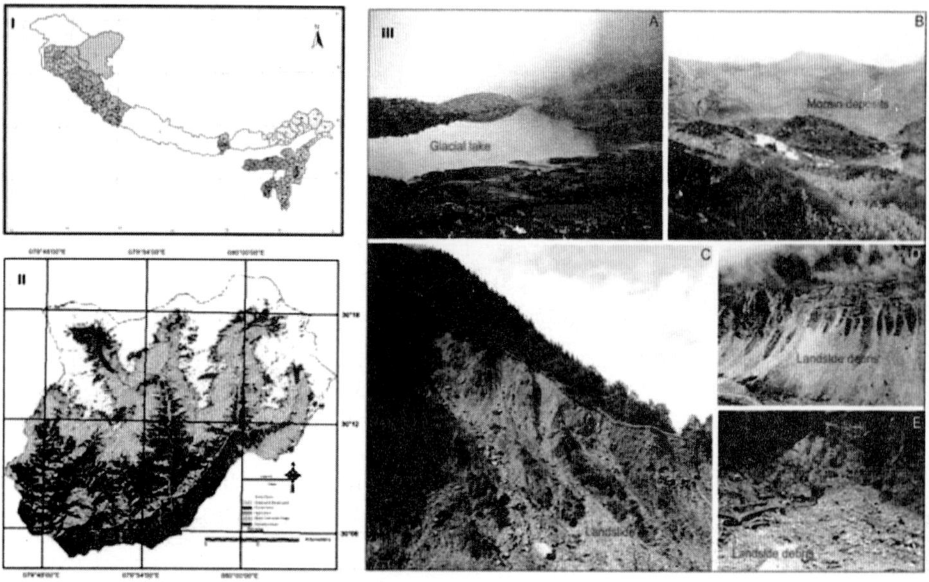

Fig. 1. Indian HImalayan Region, II: Land use Land cover Map, III: (A) Glacial lake, (B) Moraines and (C, D, E) Landslides

General features of the study area: I. location of Indian Himalayan Region. II. Land use Land cover Map prepared from Satellite data using maximum likelihood algorithm in ERDAS 2011 III. Photographs of glacial lake in the study region, moraine deposit in timber line, near a stream in the study area that may dam the stream flow into a lake, regions of landslide locations and landslide debris in the study area.

On the basis of drainage order, the Channels were classified according to the Strahler scheme (Strahler, 1964). Morphometric parameters (Linear, Arcal and Relief aspect) such as perimeter, basin area, basin length, stream numbers, stream order, and stream length were calculated. These parameters were used to determine the influence factors such as bifurcation ratio, stream length ratio, drainage density, frequency, texture, length of overland flow, form factor, circulatory ratio, elongation ratio, shape index, relief ratio, and gradient ratio. Stream Junction density, its buffer and dominant flow direction were analyzed for hydrological structure and land mass movement. Correlations between related features were analyzed by using SPSS 19 package to evaluate the role of drainage in the landform development. Morphometric features were integrated with Remote Sensing and GIS data sets to demonstrate drainage impact on the basin .

Results and discussion

Present study consist the Multicriteria Decision analysis (Strahler, 1964) to understand the morphometry of catchment area and to identify the vulnerable zones of the specific regions. Thematic maps viz. digital elevation model, relative relief map, slope map, aspects map, drainage density, stream frequency, fracture density were developed for the study. Weight and score for each thematic layer and their class has been estimated and applied for the study. Morphometric parameters has been estimated and utilised to understand the hydrological behaviour and topography of the study area.

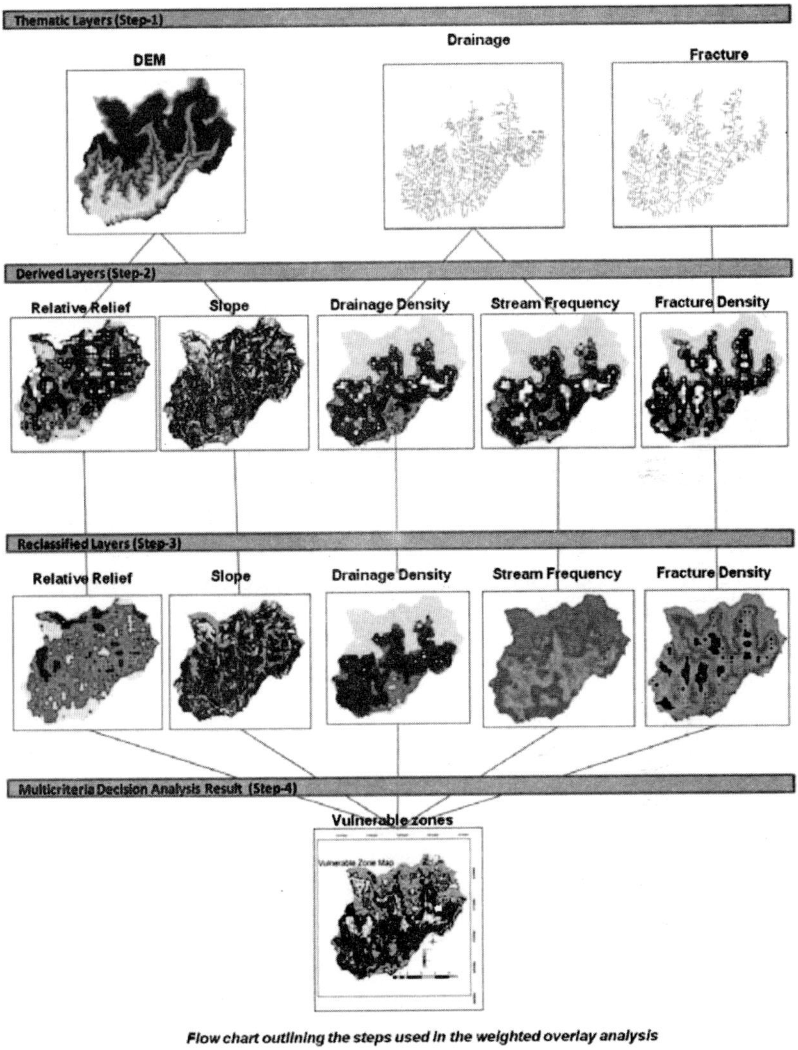

Fig. 2. Flow chart outlining the steps used in the weighted overlay analysis Schematically the steps followed in Multicriteria decision analysis (MCDA) to identify the vulnerable zones.

In Step 1: Digital Elevation Model (DEM) and drainage were derived from Toposheet; Fracture from Satellite data; and Geology from geological map.

In step 2: Relative relief, slope, drainage density, stream frequency, fracture density were derived from thematic layers of step 1.

In step 3: Above each derived layers are reclassified into five classes (very low, low, medium, high, very high)

In Step 4: Each of the above layers were associated with a weight and score and these layers were expressed in a GIS overlay analysis to derive the vulnerable zone. The vulnerable zone is presented according to degree of intensity.

The vulnerable zone map (Fig. 2) of the area is developed with the merger of different thematic layers (Slope, relative relief, drainage density, frequency, fracture density, geology and aspect). Total sum of the weights and scores of the thematic layers is considered as 100%. The fractures and the drainage network had derived the spatial distribution of the vulnerable zone. Above features are in correlation with Land use land cover (LULC) maps preferred from satellite data. Vulnerable zone are identified in the present study and the degree of vulnerable zone map developed as per the intensity of vulnerability. Major part of the basin area is prone to landslide. Among the 557.63 km^2, an area of 11.17 km^2 is very high vulnerable. High moderate vulnerable zone consists 30.5% and medium vulnerable is 40% of the basin area.

Many glacial lakes are located in the glaciers, which fall under the study area close to the area having medium drainage density. Glacial melting may cause swelling of these glacial lake that may forms supra glacial lakes. The knowledge of the location and the characteristics of these lakes are primary parameters to estimate the risk related to supra glacial lake and glacial lake outburst flood (GLOF). GLOF can have potential risk of landslides and flooding downhill area and is highly vulnerable to the fluctuations in runoff, flood and sedimentation. The overlapping threats from GLOF and landslide will cause multiple hazardous in the study area. Debris and moraines had been identified in the study area (Fig. 1). Glacier moraine consists more than 10 percentage of the study area (53. 34 km^2). Rock material and the moraine together block the drainage and act as natural dams. These dams genesis lakes in this region (Fig. 1). Lakes may form supra glacial lakes and results in GLOF. The present findings support the observation of Oliver Korup *et al.* (2010) that glacier dynamics and landslides have been argued to retard head ward fluvial erosion into the plateau interior by forming dams and protective alluvial fill.

Present study area (557.63 km^2) is part of the Nanda Devi Biosphere (more than 6000 km^2). The study area consists 12 forest types classified as per the height within an area of 425.42 km^2. The entire area of the above forest types is in landslide prone area. Majority of the area had been identified as located at high vulnerable and vulnerable area in the present study (Fig. 2). Glacial retreat and glacial melting may have catastrophic effect on Nanda Devi Biosphere Reserve.

GLOF outburst flood in Alps Mountain consist surges with high sediments load. In the present study area it may cause actual debris flow and may effect on the identified vulnerable zones of this area. These vulnerable zones may have catastrophic phenomena like glaciers lake outburst flood. The glacial lake outburst flood may further increase the disastrous landslides events. The identified locations of the above multiple hazard area is delineated in the Study area (Fig. 1). The path of the glacial outburst and the landslides will be through the region of maximum drainage density. The area estimated for respective intervals of slope shows that majority of the area of the study location falls in extreme slope . Drainage density is also high in these extreme slopes. Fracture density and drainage frequency are maximum within an area of 2 km^2. This is area of extreme vulnerable zone in the total study area of 557.63 km^2. The 'maximum vulnerable zones' are spread all over the study region. Hence any eventuality in this vulnerable zone may be drastic to the whole study area of 557.63 km^2. This may have multiple effect if there is a increased flow through drainages due to glacial melt, retreat or GLOF (Fig. 1).

Acknowledgement

The first author is grateful to Director, School of Marine Sciences for support, the second author is grateful to Director, NISCAIR for facilities and the third author is thankful to Director, CRRI for support and permission for the present study. The technical support from Mr Nasib, Project Assistant, NISCAIR is thankfully acknowledged.

Referances

Burbank D W (2005), Earth science: Cracking the Himalaya. Nature 434, 963-964

Brozović N, Burbank D W and Meigs (1997), Climatic Limits on Landscape Development in the Northwestern Himalaya. Science Vol. 276 no. 5312 pp. 571-574

Bali R, Agarwal K K, Ali N S, Rastogi S K and Krishna K, (2011), Drainage Morphometry of Himalaya Glacio-fluvial basin, India: hydrologic and neotectonic implications. Environment Earth Science DOI10.1007/s12665-011-1324-1.

Hewitt K (1988), Catastrophic Landslide Deposits in the Karakoram Himalaya. Science 64-67.

Huntington K W, Blythe A N and Hodges K V (2006), Climate change and Late Pliocene acceleration of erosion in the Himalaya, Earth and Planetary. Science Letters 252, 107–118.

Jianchu X, Grumbine R E, Shrestha A, Eriksson M, Yang X, Wang Y and Wilkes A (2009), The Melting Himalaya: Cascading Effect of Climate Change on Water, Biodiversity, and livelihoods. Conservation Biology, Vol 23, No.3, 520-530.

Jiawen R, Dahe Q, Shichang K, Shugui, Jianchen P and Zhefan J (2004), Glacier variations and climate warming and drying in the central Himalayas. Chinese Science Bulletin Vol. 49 No.1 65 69.

Korup O, Montgomery D and Hewitt K (2010), Glaciers and Landslide feedbacks to topographic relief in the Himalayan Syntaxes. PNAS vol. 107 no 12 5317-5322

Kothawale D R and Kumar R (2005), on the recent changes in surface temperature trends over India, Geophysical research letters vol 32 L18714 doi:10.1029/2005GL023528.

Kumar H and Mehrotra A (2011), Geotechnical assessment of landslides in the Bageshwar district of Uttarakhand, India. Applied Geoinformatics for Society and Environment, Germany March 12–14.

Kulkarni A V, Bahuguna I M, Rathore B P, Singh S K, Randhawa S S, Sood R K and Sunil Dhar (2007), Glacial retreat in Himalaya using Indian Remote Sensing satellite data. Current Science, Vol. 92, NO. 1, 10.

Kireet K, Dumka R K, Miral M S Satyal G S and Pant M (2008), Estimation of retreat rate of Gangotri glaciers using rapid static and kinematic GPS survey. Current Science vol.94 No.2 25.

Kuhle M (2011), Last Glacial Maximum Glaciation (LGM/LGP) in High Asia (Tibet and surrounding Mountains). In: Encyclopedia of Snow, Ice and Glaciers. Springer 697-702.

Lehmkuhl F and Owen L A (2005), Late Quaternary glaciation of Tibet and the bordering mountains: a review. Boreas, Vol. 34, pp. 87–100.

Pradhan B, Singh R P and Buchroithner (2006), Estimation of stress and its use in evolution of landslide prone regions using remote sensing data. Advance in Space research, 37 (2006) 698-709.

Raina V K (2010), Himalayan Glaciers -A State-of-art review of Glacial studies, Glacial Retreat and Climate Change, published by GBPIHED kosi-Katarmal Almora.

Strahler, A N (1964), Quantitative geomorphology of drainage basin and channel networks. In: V.T. Chow, ed. Handbook of applied hydrology. New York: McGraw Hill Book, 4-76.

Vince G (2010), Dams for Patagonia. Science: 382-385.

Valdiya K S (2010), The Making of India, Geodynamic evolution. McMillion Publishers Ltd.

Whipple K X (2009), The influence of climate on the tectonic evolution of mountain belts. Nature Geosciences 2, 97 - 104 (2009)

Wobus C, Heimsath A, Whipple K & Hodges K (2005), Active out-of-sequence thrust faulting in the central Nepalese Himalaya. Nature 434, 1008-1011 (21 April 2005).

Sustainable ecosystem development planning in Phakot sub-watershed, Tehri Garhwal district, Uttarakhand using ikonos data

T.S. Kachhwaha, P. Kunwar, A.K. Agarwal* & Anil Kumar

Forest Resources & Ecology Division,*Water Resources Division-1
Remote Sensing Applications Centre, Lucknow 226 021, India
[E-mail- tskachhwaha@gmail.com]

Introduction

The increasing population pressure, development of infrastructure facilities, especially the road network, and construction of residential colonies have not only depleted the natural resources but have also created environmental imbalance in the State Uttarakhand. Samjwal *et al.* (2006), Thompson *et al.* (2009) and Shrestha *et al.* (2011) have reported receding of Himalayan glaciers due to global warming and climate change. Upward shift of tree line in Himalayan region has also been reported by Dubey *et al.* (2003), Panigrahy *et al.* (2010) and Singh *et al.* (2012). For the development of hilly areas on a sustainable basis, it is imperative to have information on the existing natural resources, resource limitations, physical / terrain features, climatic and ecological conditions, socio-economic profiles of the area, current practices of planning and management and the contemporary technologies to be used for the sustainability of natural resources. Some of the parameters such as change in land use pattern, vegetation cover & biomass, water resources, soils etc. are important variables to observe change in the climate. Remotely sensed satellite data, which provide reliable and cost effective information, have been used for last few decades in several studies related to natural resources (Kachhwaha, 1983, Kachhwaha, 1993, Gosh *et al.*, 1996, Hong *et al.*, 1998, Singh *et al.*, 2003, Ganzin *et al.*, 2005, Yang & Liu, 2005, Navalgund *et al.*, 2007, Nichol and Wong, 2007, Wang *et al.*, 2007and Yang, 2007).

Present study in Phakot sub-watershed of Tehri Garhwal district under the 'Bio-geo Database and Ecological modeling in Himalayan Region' is initiated under a Programme of Department of Science and Technology, Govt. of India using high resolution IKONOS satellite data. In the present study, IKONOS satellite images were interpreted to prepare various resources maps. Database was spatially integrated based on certain decision rules to generate management plans for sustainable land resource development. Murthy *et al.* (1997), Verma *et al.* (1999), Kachhwaha *et al.* (2000) and Kunwar and Kachhwaha (2001) have also adopted integrated approach using remote sensing and GIS techniques to suggest site-specific action plan for sustainable development.

Materials and methods

Phakot sub-watershed forms the part of Huini watershed, which has been codified as 2B5D1 in 'Watershed Atlas of India' by the All India Soil and Land Use Survey Organisation, New Delhi (Anonymous, 1990). The study area is located in Narendra Nagar block of Tehri Garhwal district in Uttarakhand state of India at a distance of about 37 km from Rishikesh on Rishikesh-Tehri-Uttarkashi highway (Fig. 1). It lies within 30° 19' 57ë to 30° 22′ 20ë N latitude and 78° 16′ 52ë to 78° 21′ 50ë E longitude and occupies an area of about 1466 ha. Thirteen villages viz., Agar, Bhaingarki, Chilogi, Janglet, Kasmoli, Pata (Phakot), Katkor, Bhaintan, Hadisera, Malas, Naur, Lamyali and Tachhla forms the part of the study area.

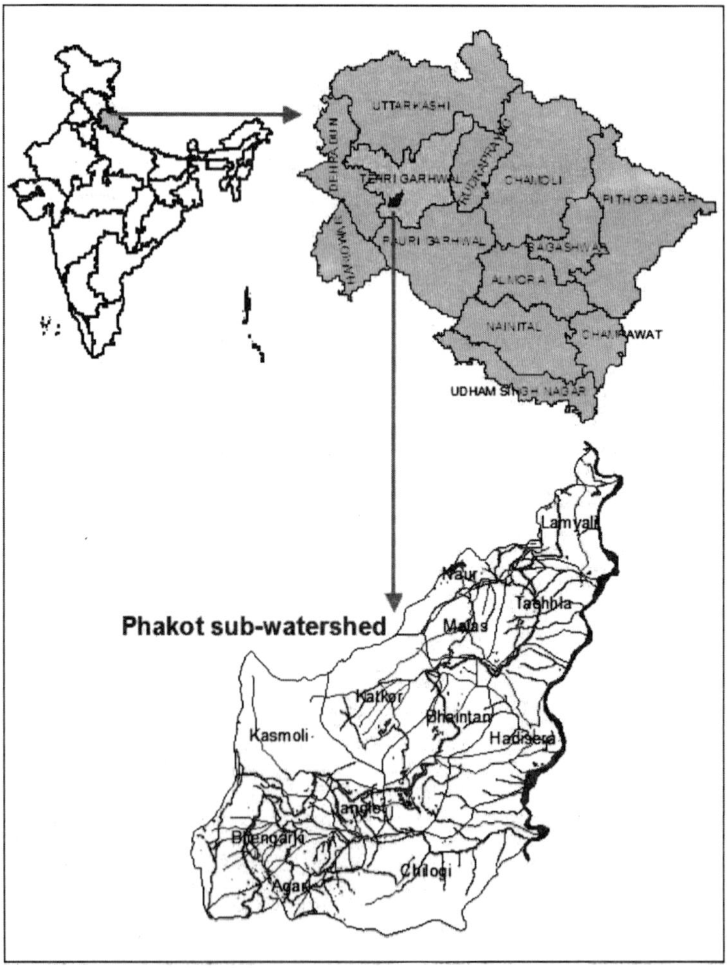

Fig. 1. Location map of the study area

The IKONOS images of three seasons *i.e.* rabi, kharif and summer, acquired on February, 2004, October, 2004 and May, 2004 respectively, were used for the present study. Topographic maps and a GPS receiver were used to collect Ground Control Points (GCPs) in order to prepare the base map. Satellite images on 1:4000 scales were visually interpreted based on standard image characteristics and using ground information to prepare thematic maps *viz.*, land use, soils, land capability, hydrogeomorphology, surface water bodies, drainage, watershed, slope, aspect etc. February image was used to identity winter season (rabi) crops. October image was used to identify monsoon season (kharif) crops. May image was used for summer season (zaid) crop identification, soil and hydrogeomorphological mapping. The IKONOS images were also used to prepare infrastructural maps of roads and settlements and for updating the drainage maps prepared form SOI topographical maps. Field data on floral and faunal biodiversity were generated by Garhwal University and Aligarh Muslim University respectively. Data related to demography and socioeconomic aspects were collected by Banaras Hindu University as participating organisations in this programme. Data provided by aforesaid participating organisations were also integrated in the digital data base.

Data thus generated was spatially integrated in GIS based on the decision rules to generate management plans for sustainable land and water resource development in the study area. Digital data base in GIS platform created under the present study comprises natural resources, suggestive measures for its management and socioeconomic profile of the area.

Results and discussion

In the study area, 519.73 ha of land is covered by dense forest, 210.61 ha by open forest, 64.43 ha by degraded forest, 92.23 ha by dense scrub and 60.98 ha by open scrub which is 35.5 per cent, 14.4 per cent, 4.4 per cent, 6.3 per cent and 4.2 percent respectively of the study area. Thus, the total area under forest in the study area, as estimated using IKONOS data, is 948 ha, which is 64.8 per cent of the total study area. Forests were further classified into evergreen and deciduous forests. Spatial distribution of various categories of land use/land cover is given in Fig. 2. Saxena (2002), Fornes *et al.* (2006) and Jain (2007) have also used high-resolution data for land cover and forest resources mapping. Land of 38.31 ha were found to be occupied by open land without scrub which is defined as wasteland (Anonymous, 1991 and Anonymous, 2006). Area occupied by various land use/land cover classes is given in Table 1. Based on interpretation of satellite data and soil sample analysis, soils of the study area were classified into 11 dominant soil series. Most of the soils of the area belong to fine loamy, mixed, mesic, Mollic Hapludalfs; fine loamy, mixed, mesic, Mollic Dystric Eutrochrepts and fine loamy, mixed, mesic, Dystric Eutrochrepts soil families. Geomorphologically, the area was categorised into 16 classes. Most of the study area falls under three major categories *viz.*, dissected hills, hill terraces and hill slopes (Fig. 3). Slope in the study area was categorized into 9 categories. About 90 percent of the study area falls under >70 per cent, 50-70 per cent and 35-50 per cent slope classes. Phakot sub-watershed was further classified into four classes with nano-micro watershed being the smallest unit (<10 ha). Each nano-micro watershed was assigned a 10-digit unique code. Based on physical and chemical parameters of soils and landscape characteristics, land capability maps were prepared (Fig. 4). In the study area, five land capability classes i.e. I, II, IV, VI, VIII have been classified which show increasing limitations for their use. Class I lands have minimum limitations for their use while class VIII lands have maximum limitations and are suitable only for wild life preservation and recreations.

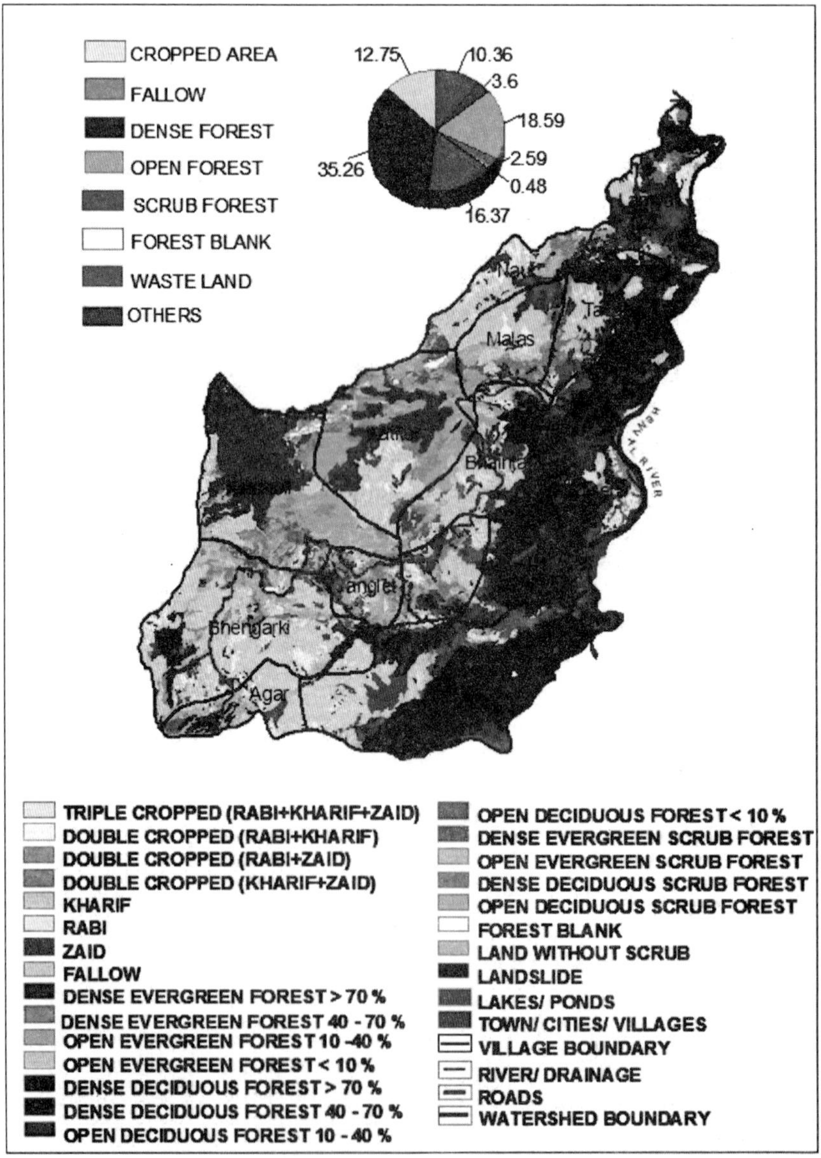

Fig. 2. Landuse/ land cover map of Phakot sub-watershed

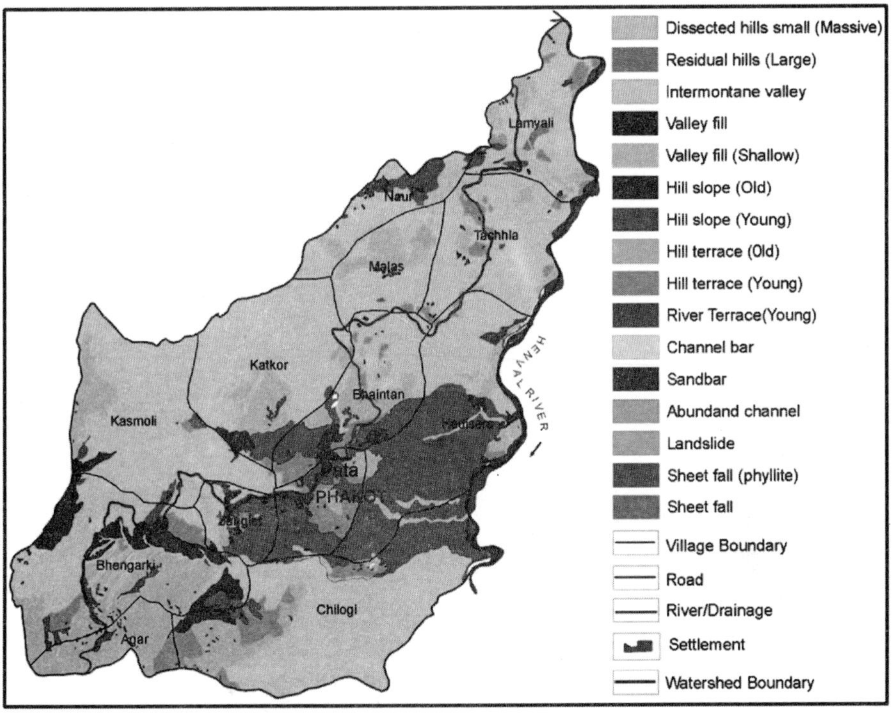

Fig. 3. Hydrogeomorphology map of Phakot sub-watershed

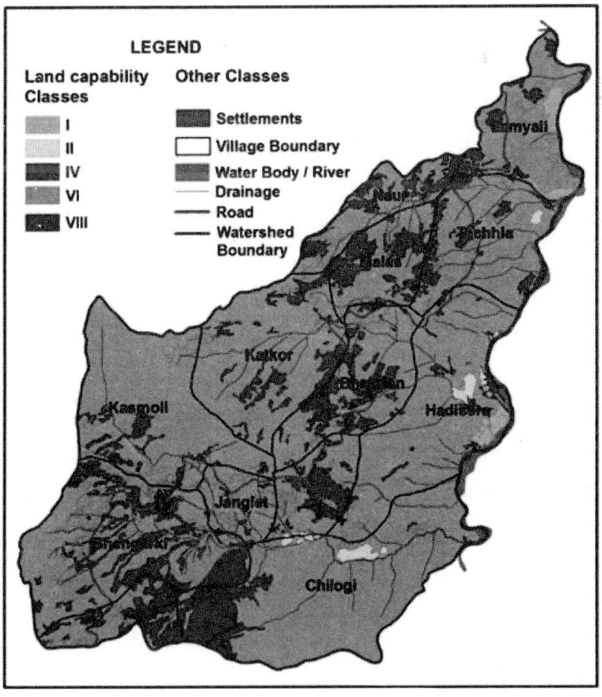

Fig. 4. Land Capability map of Phakot sub-watershed

Table 1. Area under different landuse/land cover categories in Phakot sub-watershed

Sl.No	Category	Area (ha)		
		Inside Reserved Forest Boundary	Outside Reserved Forest Boundary	Total
1.	Built-up-land	0.74	15.59	16.33
	1.1 Towns / Cities	-	0.60	0.60
	1.2 Villages	0.74	14.99	15.73
2.	Agricultural land	17.50	413.70	431.20
	2.1 Crop land	13.13	175.59	188.72
	2.1.1 Kharif	4.10	70.13	74.23
	2.1.2 Rabi	7.63	57.68	65.31
	2.1.3 Zaid	0.08	28.74	28.82
	2.1.4 Double-cropped (kharif+zaid)	0.27	4.62	4.89
	2.1.5 Double-cropped (rabi+kharif)	0.38	12.91	13.29
	2.1.6 Double-cropped (rabi+zaid)	0.59	1.34	1.93
	2.1.7 Triple-cropped (rabi+kharif+zaid)	0.08	0.17	0.25
	2.2 Fallow	4.37	238.11	242.48
3.	Forests	495.91	459.24	955.15
	3.1 Evergreen forest	116.54	120.43	236.97
	3.1.1 Dense Evergreen Forest >70%	78.02	36.77	114.79
	3.1.2 Dense Evergreen Forest 40-70%	16.12	22.6	38.72
	3.1.3 Open Evergreen Forest 10-40%	19.60	51.08	70.68
	3.1.4 Open Evergreen Forest <10%	2.80	9.98	12.78
	3.2 Deciduous Forest	352.97	204.83	557.80
	3.2.1 Dense Deciduous Forest >70%	114.32	26.27	140.59
	3.2.2 Dense Deciduous Forest 40-70%	154.35	71.28	225.63
	3.2.3 Open Deciduous Forest 10-40%	63.30	76.63	139.93
	3.2.4 Open Deciduous Forest <10%	21.00	30.65	51.65
	3.3 Scrub forest	19.23	133.98	153.21
	3.3.1 Dense Evergreen Scrub	6.82	30.05	36.87
	3.3.2 Open Evergreen Scrub	2.88	16.96	19.84
	3.3.3 Dense Deciduous Scrub	6.71	48.65	55.36
	3.3.4 Open Deciduous Scrub	2.82	38.32	41.14
	3.4 Forest blank	7.17	-	7.17
4.	Wastelands	-	38.31	38.31
	4.1 Land without scrub	-	38.31	38.31

5.	Water bodies	7.26	2.92	10.18
	5.1 River/stream	7.26	2.86	10.12
	5.2 Lakes/ponds	-	0.06	0.06
6.	Others	2.37	12.46	14.83
	6.1 Landslide	1.53	3.56	5.09
	6.2 Roads	0.84	8.90	9.74
	Total Area	523.78	942.22	1466.00

Digital database thus generated in GIS environment was integrated. All the sub-optimally utilized and wasteland areas were then identified. The management plan for sustainable land and water resources development were prepared by adopting appropriate decision rules which are based on discussions with a team of multi disciplinary scientists, district officials and local people. It was also verified in the field. These management plans recommend suitable measures to be adopted in the study area. Figure 5 shows spatial distribution of various actions to be taken up for sustainable land resources development in the study area. Table 2 shows the area to be taken for development in the study area for different type of actions. Study indicated that about 889.5 ha area has been found to be sub-optimally utilized or laying waste for which suggestive measures have been recommended to utilize this land to its full potential. Thus, about 60.67% of total geographical area of the watershed has to be taken up for one or other developmental activities. Land resources management plan suggested six types of activities based on existing potential of the land. Double cropping was suggested on the land, which is well terraced and currently used for only single crop or lying fallow and has land capability of class I and II. Agro-horticulture was suggested on the land which has <15% slope, poor land capability class (class IV or above), situated on N, NW, W and NE aspects and currently used for only single crop or lying fallow. In such areas suitable horticulture plantation could be raised along the terraces and continue to use the land for existing single crop. Agro-horticulture was also suggested for such land which has <15% slope, situated on N, NW, W and NE aspects and currently with or without open scrub but has land capability of class I and II. Agro-forestry practice was suggested in similar type of land and land use as that of agro horticulture, but such land is situated on S, SE, E and SW aspects and has relatively low moisture contents. In such land, low height local forest species could be grown along with existing land use practice. Activity of horticulture was suggested on the land with or without open scrub having poor land capability (not suitable for agriculture) and situated on N, NW, W and SE aspects. Afforestation was suggested on similar type of land as that of horticulture but situated on S, SE, SW and E aspects. Gap filling by plantation of suitable forest species was suggested on the land having open forest with crown density 10 to 40%. Table 3 indicates various management activities suggested and the decision rules applied to generate the land resource management plan.

Table 2. Land resources management plan for Phakot sub-watershed

Category	Area (ha)
Double-cropped	13.99
Agro-horticulture	151.48
Agro-forestry	245.35
Horticulture	3.82
Afforestation	259.31
Gap filling/forest protection	210.46
Landslide mitigation	5.09
No action	576.50
Total	1466.00

Table 3. Decision rules for land resources management action plan.

1. Double-cropping	
Land use	- Single crop and fallow land (existing agricultural land)
Land capability	- Class I & II
2. Agro-horticulture	
(i) Land use	- Single crop and fallow land
Land capability	- Class IV & above
Aspect	- N, NW, W & NE
(ii) Land use	- Open scrub and land without scrub
Land capability	- Class I & II
Slope	- < 15 %
Aspect	- N, NW, W & NE
Location	- Near to road & settlement
3. Agro-forestry	
(i) Land use	- Single crop and fallow land
Land capability	- Class IV & above
Aspect	- S, SE, E & SW
(ii) Land use	- Open scrub and land without scrub
Land capability	- Class I & II
Slope	- < 15 %
Aspect	- S, SE, E & SW
4. Horticulture	
Land use	- Open scrub and land without scrub
Land capability	- Class IV & above
Aspect	- N, NW, W & NE
Location	- Near to road & settlement

5. **Afforestation**

   Land use - - Open scrub, land without scrub, open forest and forest blanks

   Land capability - Class IV & above

   Aspect - S, SE, SW & E

6. **Gap Filling/Forest Protection**

   Land use - Open forest

7. **No Action**

   Land use - Settlements, triple-cropped area, double-cropped area, dense forest, forest/ horticulture plantations, water bodies, rivers/streams and roads

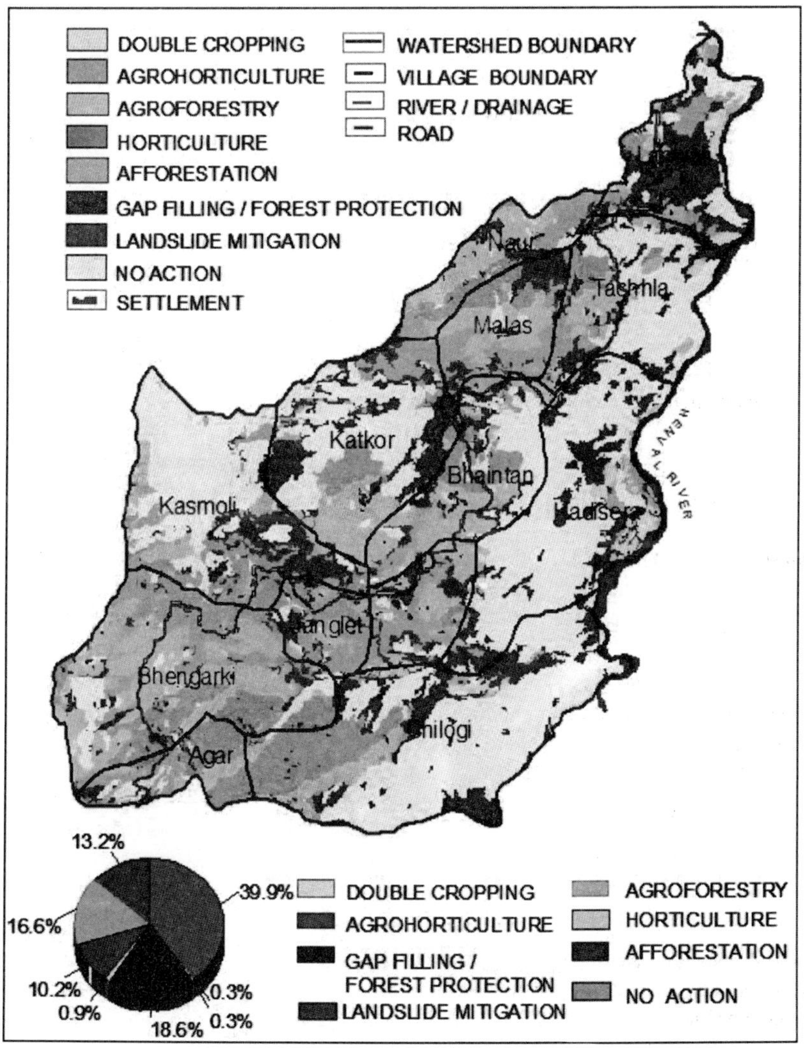

Fig. 5. Land resources management action plan

For optimal utilization of available water resources and harvesting rain water in the study area for the purpose of implementation of land resources management plan, soil conservation to check soil erosion, land degradation and landslides, and for drinking water supply, a water resources management plan was also prepared (Fig. 6). Water resources management plan includes construction of nala bunds, diversion gates, contour canals, storage tanks, dam, lift irrigation points, surface pond/contour trenches and laying of drinking water pipe lines and underground pipe lines. Details of the water resources management plan are presented in Table 4.

Table 4. Decision rules for generation of water resources management action plan in Phakot sub-watershed

Sl. No.	Water Resources Management Plan	Villages	Land Resources Manage-ment Plan	Water availability/ Source	Geomo-rphology	Remarks
1.	Diversion gate	Bhaingarki, Chilogi, Tachhla	Agro-horticulture	Streams	Hill slope, hill terrace	The diversion gates have been suggested on 1st and 2nd order streams to divert the water for storage and supply elsewhere depending upon requirements. This will be a open/close gate type system.
2.	Lift irrigation point	Chilogi	Agro-horticulture	from the river	Hill slope, hill terrace	The water will be lifted from 2nd and 3rd order stream for the supply of irrigation water to Chilogi village.
3	Dam site	Bhaingarki, Chilogi	Gap filling/ forest prot-ection/ no action	River	Valley fills, hill slope, & denudational hill	3rd order and 4th order streams have been cho-osen in almost flat terrain with large catchment area.
4.	Storage tank and contour canal	Lamyali, Tachhla, Malas, Bhaintan, Phakot, Bhaingarki, & Chilogi	Agro-horticulture, agro-forestry	Streams diverted through diversion gate & sup-plied thro-ugh contour canal	Hill terrace	The storage tank on individual land will store water for lean period. The canal along the terraces will carry water on the same terrace existing between two drainages.
5.	Drinking water pipe line and under-ground pipe line	Bhaingarki & its hamlets	No action	Tank, canal	Hill slope/hill terrace	The water through pipe line will supply water to villages.

6.	Surface pond/ contour trenches	Bhaingarki, Phakot, Katkor, Bhaintain, Malas & Lamyali	Double crop, Kharif, fallow	Rain Water	Hill terrace, hill slope, denudational hill	Very gentle sloping areas are suitable for conserv-ation of water, which will further recharge the springs/ streams in down streams.
7.	Nala bund sites	All villages	Forest and agricultural land	Streams	Varied geomorp-hology	The nala bunds are proposed on 1st order drainage on high slopes areas to check the water gradient.

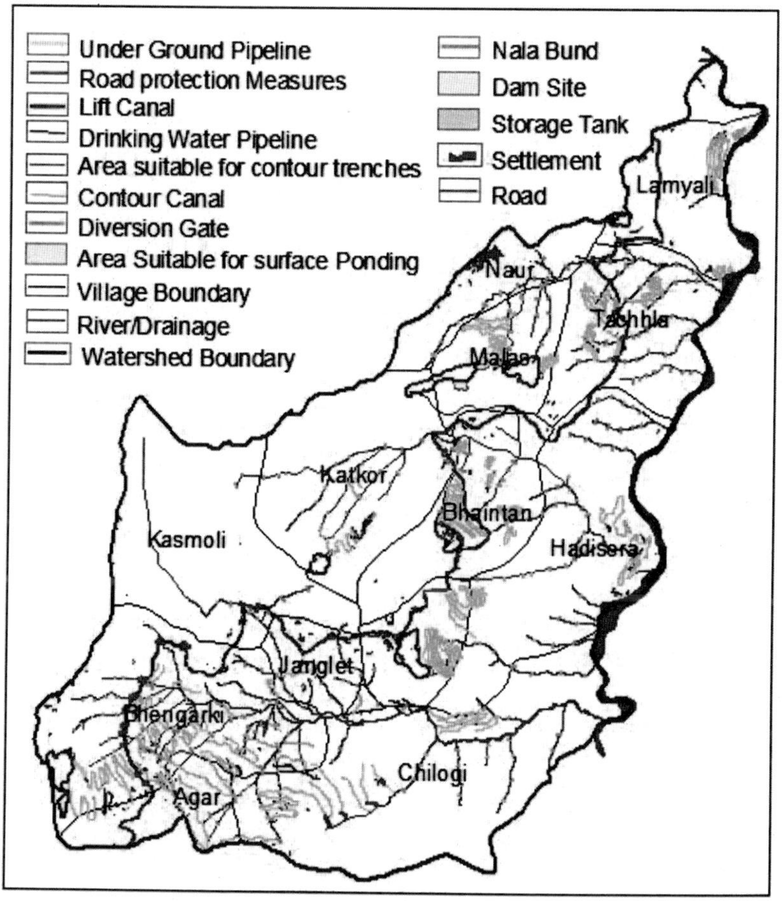

Fig. 6. Water resources management action plan

Land and water resources management plans generated in the present study are based on detailed discussions and suggestions by local people and local officers of different departments of the State Government. These management plan would help in the management and mitigation of the problems related to land degradation, scarcity of water for drinking purpose and agricultural uses.

This would lead to optimum utilization of land which is currently either lying waste or sub-optimally utilized in the Phakot sub-watershed. These recommendations would result in increase in agricultural production, upliftment of socio-economic status of the people, improved environment and thus, would pave the way for sustainable development. Data generated under the present study, many of which are important variables to understand climate change, could also be used as base line information for future studies related to climate change.

Acknowledgements

Authors are thankful to Director, Remote Sensing Applications Centre, U.P., Lucknow for providing facilities to carryout the present study and for permission to publish this paper. Thanks are also due to Dept. of Science and Technology, Govt. of India for funding the study. The authors also acknowledge the participating organisation who have provided field data on flora, fauna and socio-economic profile of the area for integration in GIS data base.

References

Anonymous, *Watershed Atlas of India.* All India Soils and Land Use Survey, Ministry of Agriculture, (New Delhi, India) 1990.

Anonymous, *Description and classification of wastelands.* National Wastelands Development Board, Ministry of Environment & Forests, Govt. of India, (New Delhi, India) 1991.

Anonymous, *Manual – National land use land cover mapping using multi-temporal satellite data.* National Remote Sensing Agency, Dept. of Space, Govt. of India, (Hyderabad, India) 2006, pp. 1-125.

Bajracharya S R, Mool P K and Shrestha B R, The impact of global warming on the glaciers of the Himalaya, In: *Proceedings of the International symposium on Geo-disasters, Infrastructure Management and Protection of World Heritage Sites,* (Nepal Engineering College, Ehime College and National Society for Earthquake Technology, Nepal), 2006, pp. 231 – 242.

Dubey B, Yadav Y R, Singh J and Chaturvedi R, Upward shift of Himalayan Pine in Western Himalaya, India, *Current Science*, 85 (2003) 1135-1136.

Fornes A, Basterretxea G, Orfila A, Jordi A, Alverez A and Tintore J, Mapping Pesidonia Oceanica from IKONOS, *International J. Photogrammetry and Remote Sensing,* 60 (2006) 315 - 322.

Ganzin N, Coetzee H, Rothauge A. and Fotsing J M, Rangeland Resources assessment with satellite Imagery: An operational tool for national planning in Namibia, *J. Geocarto International,* 20 (2005) 33 - 42.

Gosh S, Sen K K, Rana U, Rao K S and Saxena K G, Application of GIS for land use/ land cover change analysis in a mountainous terrain, *Photonirvachak, J. Indian Society Remote Sensing,* 24(1996) 193-202.

Hong S He, David J M, Volker C R and Thomas R C, Integration of GIS data and classified satellite imagery for regional forest management, *Ecological Applications, Ecological Society of America,* 8(1998) 1072-1083.

Jain S, Use of IKONOS Satellite data to identify informal settlements in Dehradun, India, *International Journal of Remote Sensing,* 28(2007) 3227 - 2333.

Kachhwaha T S, Spectral signatures obtained from Landsat digital data for forest vegetation and land use mapping in India, *Photogrammetric Engineering and Remote Sensing,* 49(1983) 685-689.

Kachhwaha T S, Temporal and multisensor approach in forest vegetation mapping and corridor identification for effective management of Rajaji National Park, Uttar Pradesh, India, *International J. Remote Sensing,* 14(1993) 3105-3114.

Kachhwaha T S, Kunwar P and Agarwal A K, Application of remote sensing and GIS techniques in generation of site-specific action plans for sustainable land and water resources development, In: *Proceedings* of *All India Seminar on Environment and Social Issues in Water Resources Development, Institute of Engineers,* edited by J. Mohan, R.K. Khanna and Nirupama, (Institute of Engineers, Lucknow, India) 2000, pp 154-166.

Kunwar P and Kachhwaha T S, Transformation of land use/ land cover for optimal utilization of natural resources on sustainable basis in Mandawara block, Lalitpur district, Uttar Pradesh using remote sensing and GIS techniques, In: *Environment and Development,* edited by S.B. Singh and Rana P.B. Singh, (National Geographical Society of India, Banaras Hindu University, Varanasi, India), 2001, pp. 151-156.

Murthy Y V S, Uday Bhasker N, Kameswara Rao S V C and Perumal A, Geographical information systems approach to suggest land use alternative for sustainable rural development, In: *Proceedings of ICORG on Geographical Information Systems and Remote Sensing Applications*, (Hyderabad, India) 1997, pp. 44-47.

Navalgund R R, Jayaraman V and Roy P S, Remote sensing applications: An overview, *Current Science,* 23 (2007) 1747 - 1766.

Nichol J and Wong M S, Remote sensing of urban vegetation life form by spectral mixture analysis of high-resolution IKONOS satellite images. *International J. Remote Sensing,* 28(2007) 985 -1000.

Panigrahy S, Anitha D, Kimothi M M and Singh S P, Timberline change detection using topographic map and satellite imagery. *Tropical Ecology, 51*(2010) 87-91.

Saxena A K, Forest type stratification with emphasis on species and density in Timli Badkala forest range using high resolution Indian Remote Sensing satellite data In: *Proceedings of ISPRS Commission VII Symp. Resource and Environmental Monitoring*, edited by R.R. Navalgund, S.R. Nayak, R. Sudarshana, R. Nagaraja and S. Ravindran, (National Remote Sensing Agency, Hydrabad, India) 2002, pp. 1036–1048.

Shrestha A B and Aryal R, Climate change in Nepal and its impact on the Himalayan glaciers, *Regional Environmental Change,* 11 (2011) 65-77.

Singh I J, Das K K and Kushwaha S P S, Forest stock assessment using IRS LISS III and PAN merged data in Timili forest range, Dehradun, *Photonirvachak, J.Indian Society Remote Sensing*, 31(2003) 49-56.

Singh C P, Panigrahy S, Thapliyal A, Kimothi M M, Soni P and Parihar J S, Monitoring the alpine tree line shift in parts of the Indian Himalayas using remote sensing, *Current Science*, 102(2012) 559-562.

Thompson L G, Brecher H H, Mosley-Thompson E, Hardy D R and Mark B G, Glacier loss on Kilimanjaro continues unabated, *Proceedings of the National Academy of Sciences,* 106 (2009) 19770–19775.

Verma V K, Sharma P K, Patel L B, Loshali D C and Toor G S, Natural resource management for sustainable development using remote sensing (Abstract), In: *Pre-Conference Proceedings of Map India* (New Delhi, India) 1999, pp. 97-100.

Wang X, Ren Z and Tan K, Eco-environmental evaluation with remote sensing in middle and upper areas of the Yellow River drainage basin, *International J. Remote Sensing,* 28(2007) 3937 - 3951.

Yang X, Integrated use of remote sensing and geographical information system in riparian vegetation delineation and mapping, *International J. Remote Sensing*, 28(2007) 353 - 370.

Yang X and Liu Z, Using satellite imagery and GIS for land use and land cover change mapping in an Esturian watershed, *International J. Remote Sensing,* 26(2005) 5275 - 5296.

Consequences of climate change – Risks to the globe

Meen B. Poudyal Chhetri

[1]Disaster Preparedness Network–Nepal (DPNet) and Nepal Center for Disaster Management (NCDM), Nepal
[E-mail: dr_mbpchhetri@hotmail.com]

Introduction

Climate change is one of the key factors for the occurrences of various types of natural disasters in Nepal. The vast altitudinal variation within a short span of about 193 km, ranging from 60 meters to 8848 meters above sea level makes the country an abundant storehouse of bio-diversity and ecological niches with diverse agro-climatic zones ranging from the sub-tropical to the alpine and tundra (Poudyal Chhetri and Bhattarai 2001). On the other hand, increasing population, rapid and unplanned urbanization and other economic activities in vulnerable areas are other contributing factors to increase hazards. Hence, Nepal is a global hot spot for several types of disasters.

The monsoon of South Asia is one of the most important and influential phenomena of the earth's climate and eco system. The Tibetan Plateau warms rapidly relative to the Indian Ocean during summer in the Northern Hemisphere. The resulting low pressure over Asia and higher pressure over the ocean gives rise to the strong low-level atmospheric pressure gradient that, in turn, generates the south-west monsoon from the Indian Ocean. In years of low snowfall, the Tibetan Plateau is able to warm earlier and generate stronger monsoonal circulation. Deep snow and associated influences on albedo (reflected light or radiation) and soil hydrology delay and weaken the monsoon. South Asian monsoon during the last glacial period was significantly weaker than at present, and was abruptly strengthened during the beginning of Holocene Period around 12,000 years ago (Overpeck and Cole 2007).

Materials and methods

Global warming, climate change issues and the mountainous regions

Various studies have shown that in the last 100 years the world's average temperature has risen rapidly. For the last 10,000 years we have been living in a remarkably stable climate that has allowed the whole of human development to take place, now we see the potential for sudden changes of between 2 and 6 degrees Celsius (by the end of this century). We just do not know what the world is like at those temperatures. We are climbing out of the safe zone (Corell 2007).

The scale of temperature rise is in increasing trend. Out of the 10 recorded warmest years in history, nine were recorded during the last decade. The global mean temperature is expected to increase between 1.4 to 5.8°C over the next hundred years. The adverse effects of such change in global climate are seen in the Himalayas where glaciers and glacial lakes are posing catastrophic risks. Himalayan glaciers are retreating at rates ranging from 10 to 60m per year and many small glaciers (<0.2 sq.km) have vanished. The boundary of most of the high altitude valley glaciers in Bhutan, China and Nepal are diminishing quickly. Glaciers in the Himalaya are thinning faster than elsewhere in the world. If current situation prevailed, the glaciers could disappear by the year 2305. Thus, climate change is shrinking the mountain glacier and directly affecting the landscape and threatening water supplies all over the world. In such a way, the Himalayan glaciers can be considered as a reliable indicator of climate change.

Mountain regions occupy about a quarter of the global terrestrial land surface and provide goods and services to more than half of the inhabitants. The rise of the Himalaya and Tibetan Plateau together that started about 50 million years before caused a tremendous impact on the regional and global climate of the world. Himalayan region has long been recognized as extremely rich in animal and especially plant diversity. Himalayan watersheds harbor more diverse ecosystems than the Amazon. The rise of the Himalaya and Tibetan Plateau together caused a tremendous impact on the regional and global climate of the world. During the summer season warm moist wind blows from ocean to land. While in winter time, cold dry wind blows from land to ocean. Monsoon circulation involves a change of approximately 180 degree in the direction of wind between the summer and winter (Upreti, 2008).

Nepal's 83% land mass is mountainous terrain. The wide range in altitudinal variation along its width gives rise to a steep and rugged topography and extreme relief. Steep and unstable slopes, rugged terrain, active geodynamic processes and intense monsoon rains make the Himalaya an active and fragile mountain range. As the nature of the Himalaya suggests, landslides and debris flows and floods are the main types of water-induced hazards in the region and in Nepal. These hazards wipe out entire villages, wash out roads, bridges, canals and hydropower plants and damage hectares of valuable agricultural land during the monsoon season. Besides substantial economic losses, more than 320 people on average lose their lives in the Nepalese Himalayan region alone. Other losses from these hazards are on a rise every year.

Many factors trigger debris mass movement or debris flows. Among the most common triggers in the Himalaya are prolonged or heavy monsoon rains. Rainfall can lead to mass movement of debris by reducing the internal or binding strength of soil and other materials through three different mechanisms. The saturation of soil materials increases the weight of slope materials and creates greater gravitational force. Saturation of soil materials can reduce the cohesive bond of individual soil particles and water can serve as a lubricant along the interface between soil and rock and along the weakness zones of rocks, such as joints, cracks and fault planes.

The intensity and duration of rainfall that can initiate a landslide depends on many factors. However, most landslides and debris flows reported in the Nepal Himalaya are either associated with intense or sustained monsoon precipitation (Dhital 2003, Adhikari and Koshimizu 2005).

Geo-scientists have found that due to global warming, the number and volume of glacier-lake outburst flood hazards are on the rise. Some of these floods have produced discharge rates of up to 30,000 m^3/sec and can run for distances of 200 km (Richardson and Reynolds, 2000).

Considering the average vertical lapse rate of 6.5°C per kilometer, it was found that almost 20% of the present glaciated area above 5000 meter altitude are likely to be snow and glacier free area with an increase of air temperature by 1°C. Similarly, 3°C and 4°C rise in temperature could result into the loss of 58% and 70% of snow and glaciated areas respectively. Such changes are likely to contribute to the faster development of glacier lakes leading consequently to the increase in potential of glacier-lake outburst flood hazards. Also, increase in precipitation by more than 20% is likely to cause significant increase in sediment delivery and more than 20% increase in annual sediment deposit could be expected in a scenario of 50% increase in annual precipitation (MoPE, 2004).

The above figures and situation warn us on the effects of global warming and climate change in the GLOFs. The outburst of huge GLOFs may cause enormous loss not only in Nepal that will extend up to North India, Bhutan and China as well.

Results and discussion

Effects of the green house gases (GHG)

Global climate change –driven largely by human induced warming of greenhouse gases (GHG) – is a growing threat to humanity. The world experienced a surface temperature rise of 0.6°C on average during the 20th century, and the temperature by year 2100 is projected to go as high as 6.4°C relative to 1990 if GHG emissions are not reduced (IPCC 2007). But time has shown that given the dependence of global economic systems on fossil fuels, and the time required for new technologies that reduce or replace fossil fuels, to integrate into the global marketplace, significant reduction in GHG emissions is unlikely to occur soon enough to avoid climate impacts. Significant harm from observed climate change on the environment and on society is already occurring worldwide and more severe and widespread impacts lie ahead. Climate change impacts on the geo-environment of the Nepal Himalaya are significant.

Implications of climate change in the Nepalese Himalayan region & beyond

Although Nepal is not contributing to global warming, it has tremendously impacted the glacier ecosystem in the Nepalese Himalayas. Climate change is not just an environmental phenomenon but also an economic, social and political issue in Nepal. From the point of view of climate change, Nepal is among the most vulnerable countries in the world. Himalayan Study is important because of the four reasons, namely; role of Himalaya on global climate, present impact of global warming in the water balance in high mountain areas, the impact of climate change in flora and fauna and the socio-economic impact of climate change in subsistence farming.

The major impacts of climate change in Nepal are: increased glacier-lake outburst flood (GLOF) hazards, increased variability of river runoff, increased sediments, increased evaporation from reservoirs and impacts on watershed. As a result glacier melt and precipitation patterns would occur. Nepal has wide variety of species. A study has found that 2.4% of biodiversity may be lost with climate change. Obviously climate change will affect agriculture. While majority of the people of Nepal depend on agricultural crops like rice, maize and wheat. Higher temperatures, increased evapo-transpiration and decreased winter precipitation may result into droughts. It should be considered as an early warning for food security.

The average maximum temperature in Nepal between 1977 and 1999 has increased by 0.9°C, at a rate of 0.03°C to 0.12°C per year, whereas the global average surface temperature rise of the last century was 0.6±0.2 °C. (Shrestha *et al.* 1999) and is estimated to have gone even higher since then. This is one of the highest registered rates of temperature rise in the world. The observed trend of rising temperature in Nepal is challenging the IPCC projections, as it seems that land areas will warm more rapidly than the global average.

Almost 20% of the glaciated areas in Nepal above 5000 m are likely to be snow and glacier free area at an increase of air temperature by 1°C. Two degree Celsius rise in temperature can cause the loss of almost 40% of the areas. Similarly, 3°C and 4°C rise in temperature can result in the loss of about 58 % and 70% of snow and glacier areas, respectively.

Conclusions

Greenhouse gases from human activities are among the major causes for the alarming situations of climate change and global warming. Various studies show pressing need to calculate carbon appropriation, the basis for calculating the impact of ecological imbalance particularly - deforestation. Stopping deforest-ation and building healthy environment should be the key issues in climate change policy of every government. Then it will provide a way for millions of poor people in developing countries to benefit directly. Such policy will help to reduce deforestation, maintain ecological balance and allow the nations to sell credits for successful programs combating carbon dioxide. On the other hand, developed countries that pollute more than the allowed limits under the existing Kyoto accord would be able to buy the carbon credits to increase their emission levels and help to fund forest protection plan and programs.

References

Adhikari D P, (2003), Monitoring the Monsoon in the Himalayas: Observations in Central Nepal, Monthly Weather Review, Vol. 131, pp. 1408-1427, June, 2001.

Adhikari D P and Koshimizu S (2005), "Debris Flow Disaster in Larcha, Upper Bhotekoshi Valley, central Nepal." In The Island Arc, 14 (4): 410-423.

Bajracharya B, Shrestha A B and Rajbhandari L (2007), Glacial Lake Outburst Floods in the Sagarmatha Region: Hazard Assessment Using GIS and Hydrodynamic Modeling, Mountain Research and Development, Vol. 27, pp. 336-344, Kathmandu.

Bhattarai D and Poudyal Chhetri M B (2001), Mitigation and Management of Floods in Nepal, Ministry of Home Affairs, HMG/Nepal.

Caine N and Mool P K (1982), Landslide in the Kolphu Khola Drainage, Middle Mountains, Nepal. In Mountain Research and Development, 2: 157-67.

Corell R (2007), Chairman of the Arctic Climate Change Impact Assessment (ACCIA).

Higuchi K, Ageta Y and Yasunari T (1982), Characteristics of Precipitation During the Monsoon Season in High Mountain Areas of the Nepal Himalaya, Hydrological Aspects of Alpine and High Mountain Areas, July, International Association of Hydrological Sciences, (IAHS), Publ. No. 138.

ICIMOD (1996), Database of Key Indicators of Sustainable Mountain Development in Nepal, Mountain Environment and Resources Information Service (MENRIS), pp.85, ISBN 92-9115-516-0, Kathmandu.

ICIMOD (2007), Disaster Preparedness for Natural Hazards, Current Status in Nepal, International Center for Integrated Mountain Development (ICIMOD), Kathmandu.

IPCC (2007b), Climate Change 2007: The Physical Basis, Contribution of Working Group I to the Fourth Assessment Report of the Intergovernmental Panel on Climate Change. Cambridge: Cambridge: University Press, United Kingdom and New York, USA. URL: http://ipcc-wg1.ucar.edu/wg1/wg1-report.html.

Kansakar S, Hannah D, Gerrad J and Rees G, (2004), Spatial Patterns in the Precipitation Regime of Nepal, International Journal of Climatology, Vol. 24, pp. 1645-1659.

Lal M, Whetton P H, Pittock A B and Chakraborty B, (1998), Simulation of Present-Day Climate Over the Indian Subcontinent by General Circulation Model. In Terrestrial, Atmospheric and Oceanic Science, 9(1): 69-96

Mirza M Q (2003), Three Recent Extreme Floods in Bangladesh: A Hydro-Meteorological Analysis, In Flood Problem and Management in South Asia, Mirza, M. Q. and Dixit, A. (eds.), pp. 35-64, Kulwer Academic Publishers, 3300 AZ Dordrecht, The Netherlands.

Mool P K (1995), Glacier Lake Outburst Flood in Nepal, Journal of Nepal, Geological Society, Vol. 11, Special Issue, pp. 273-280, Kathmandu.

Ministry of Population and Environment (2004), Annual Report, Kathmandu, Nepal.

Nepal Disaster Report: The Hazardscape and Vulnerability (2009), Disaster Preparedness Network-Nepal.

Overpeck J T and Cole J E (2007), Climate Change Lesson From a Distant Monsoon. In Nature, 445: 270-271

Richardson S D and Reynolds J M (2000a), An overview of glacial hazards in the Himalayas. *Quaternary International*, 65-66, 31–47.

Roberts M and Brice (2006), While Speaking With The Media in Kathmandu.

Shrestha A B; Wake C P; Mayewski P A and Dibb J E (1999), Maximum Temperature Trends in the Himalaya and Its Vicinity: An Analysis Based on Temperature Records From Nepal for the Period 1971-94. In Journal of Climate, 12: 2775-2787.

Upreti B N (2008), Climate change and its Implication in the Himalayan region, A Paper Presented at the Queensland University of Technology, Australia.

Watanabe T, Ives I D and Hammond J E (1994), Rapid Growth of a Glacial Lake in Khumbu Himal, Himalaya, Prospects for Catastrophic Flood, Mountain Research and Development, Vol. 14, No. 4, pp. 329-340.

Yamada T (1993), Glacier Lakes and their Outburst Floods in The Nepal Himalaya, pp. 37, Published by the Water and Energy Commission Secretariat (WES), Kathmandu.

Session- II

Socio-economic Perspectives

Impacts of climate change and coping strategies by mountain communities of Uttarakhand Himalaya, India

Dinesh Pratap

Department of Geography, DAV PG College, Dehradun 248001, Uttarakhand, India
[E-mail: d.pratap56@gmail.com]

Introduction

Climate change today is an accepted fact all over the world. Intergovernmental Panel on Climate Change (IPCC) in its Fourth Assessment Report (2007) was categorical that warming of the climate system is unequivocal as evident from observations of increased average air and ocean temperature, widespread melting of snow and ice and rising average sea level (UNEP, 2012, IPCC, 2012). The 20th century's last two decades were the hottest in 400 years and possibly the warmest for several millennia. The analysis of data for more than 150 years reveals that the earth's temperature has increased by 0.74°C in last hundred years and 12 out of last 13 years are among the warmest years since 1850 (SAP, 2011).

It has been widely acknowledged that anthropogenic activities, more than anything else, have contributed towards concentration of green house gases (GHGs) in earth's atmosphere. It has been scientifically proved that increase in concentration of GHGs has led to 'warming of the earth's surface and atmosphere, threatening to change the climate and impact the entire earth system in an unprecedented manner (Shukla, Sharma *et al.*, 2002). While the opinions differ on the rate of global warming, the fact remains that it is happening and even when the questions have been raised on IPCC projected scenario, it is suggested that the warming of 21st Century will be around 1.4°C (2.5°F) (Michaels, 2010).

The impact of climate change is considered more intense in mountain areas. The climatic variability with in short horizontal distance due to altitudinal variation, rapid change in vegetation and hydrological characteristics make mountain more vulnerable to any climate change phenomena (ICIMOD, 2012). The Himalayas which are the largest high-land mass in the world, not only play important role in global climate but are themselves subject to climate change impact and warming here is predicted to be well above the global average (Pandey & Vekataraman, 2012). Though, at times, the opinions differ on impact of global warming on Himalayan glacier melt and some scientists feel that the Himalayas are quite safer zone as the conditions here are controlled by the winter snowfall rather than external factors like global warming, it has been underlined that there is a change in climate in terms of shrinking of winter period and glacial retreat (Hindustan Times, 2011). Though there was no consensus on the changes in nature of ice cover in the Himalayas among the participants of an international conference on Himalaya, they emphasized need to continue the efforts for mitigation of, and adaptation to, climate change that has been initiated by the international community (UNEP, 2011). A well documented study has indicated higher rate of warming in Himalayas than the global average and faster receding of glaciers here than in other parts of the world. The study has expressed great concern as the region has more snow and ice than any other region in the world outside the polar caps and Himalayas affect the climate of much of the South Asia (Singh, S P, 2012).

Materials and methods

Uttarakhand Himalaya region

The Physical Environment

Uttarakhand Himalayan region is located in the central Himalayan region of India extending between 28°41′ and 31°25′ North latitude and between 77°34′ and 81°4′ East longitude. It covers an area of 53,483 km^2, inhabited by 10.1 million people. Uttarakhand is a rugged mountain region where altitude varies from 300 m to 7800 m from south to north with in a horizontal distance of about 200 km. Except for a narrow belt of 10-25 km in south, the larger part of the region is made up by mountainous terrain and is divided into low Siwalik mountains/Outer Himalaya (700-1200m), the Lesser Himalaya (1200-3000m), Higher Himalaya (3000-7800m) and the Trans Himalaya (4500-5700m) (Kumar *et al.* 2011). The altitudinal variation has resulted in diversity of topography, meteorology, natural vegetation and socio-economic characteristics, truly representing the whole of Himalayan ecosystem.

Uttarakhand Himalaya exhibits spatial variation in regional and micro-climate because of altitudinal variations, slope aspects, ridges and deep longitudinal valleys, proximity to glaciers, mountain peaks, forests and other geographical factors (Rawat, 2011). Precipitation is mostly received as mon-soon rainfall from June to September, however, the greater Himalayan region experience snowfall in winter months. Annual average rainfall varies from 1250mm and 2000mm while the average annual temperature ranges from 25° Celsius in the south to sub-zero in the north. On the basis of altitude, the state can be divided into five climatic zones - (i) Tropical zone - below 900m (ii) Sub-tropical or Warm–Temperate zone - between 900m to 1800m (iii) Cool Temperate zone – between 1800m and 2400m (iv) Cold temperate zone – between 2400m and 3000m (v) Alpine and Glacial zone – above 3000m. Presence of glaciers and snow covered areas plays an important role in the ecology of the region and has made Uttarakhand Himalaya rich in hydrological resources. Besides perennial rivers, the presence of glacial lakes in the upper reaches and the fresh water lakes in the middle Himalaya form an integral part of drainage system and influence local ecology and drainage system. Some of the important glaciers of Uttarakhand such as Gangotri, Nandadevi, Bandarpunch, Chturangi, Satopanth, Khatling, Pindari, Kamet and Milam, vary in length from 5 to 29 kms.

Socio-economic characteristics

The terrain and climate of Uttarakhand have primarily guided the socio-economic characteristics of the state. Forests occupy close to 64% land area while agriculture is practiced over 13% of the geographical area. However, agriculture is main occupation of more than 70% of the working population. But typically the hill agriculture is characterized by small fragmented land holding, scattered and rain-fed fields, low productivity and limited use of modern technology. One of the highlights of agriculture–livestock based rural economy of the region is engagement of women workforce whose 90% is engaged in this sector and almost all the agricultural operations are performed by them. While the literacy rate is comparatively high at 80%, majority of the population (70%) resides in rural area. Horticulture though limited to about 4% of the cultivated area, contribute significantly through production of sub-tropical and temperate fruits and off-season vegetable. Pilgrimage and tourism are another important source of income for local people but their seasonal nature and low investment by visiting tourists does not contribute towards regular cash flow in local economy.

While the industrial activities and service sector have picked after the creation of separate state, their benefits are restricted to the southern plain and valley areas where as 90% of the region remains largely outside the influence of new developments. It is, thus, the majority of mountain community residing in more than 12000 scattered small villages in the middle and higher Himalaya, dependent on subsistence un-irrigated agriculture and limited horticultural production, which are most prone to climatic change impacting Uttarakhand Himalaya.

Results and discussion

Impact of climate change on Uttarakhand

The evidence of climate change in Uttarakhand Himalaya

Uttarakhand Himalaya which is part of Hindukush Himalayan mountain ranges, lacks in long-term detailed climatic data. Except for Dehradun (670m), Pantnagar (244m), Almora (1640m), Nainital (1950 m) and Muketshwar (2286m), the long-term reliable data are not available. Given the climatic and altitudinal variation and the area covered by state (>53000 km^2), it is extremely difficult to get concrete regional evidence of climatic change. However, the studies are available for particular locations that provide information on the temperature, rainfall, glacier melt and other parameters indicating changes in their behavior. For example, the First Interim Report of Uttarakhand Centre on Climate Change established by Government of Uttarakhand has reported that the available data of a lesser Himalayan hill station, Almora (1640m) reveals that average temperature has increased by 0.46°C in last 53 years (1955-2007) (Fig. 1) and the rainfall records of the same place and another location, Manora Peak, near Nainital (1950m) (Fig. 2) show decreasing trend (UCCC, 2011).

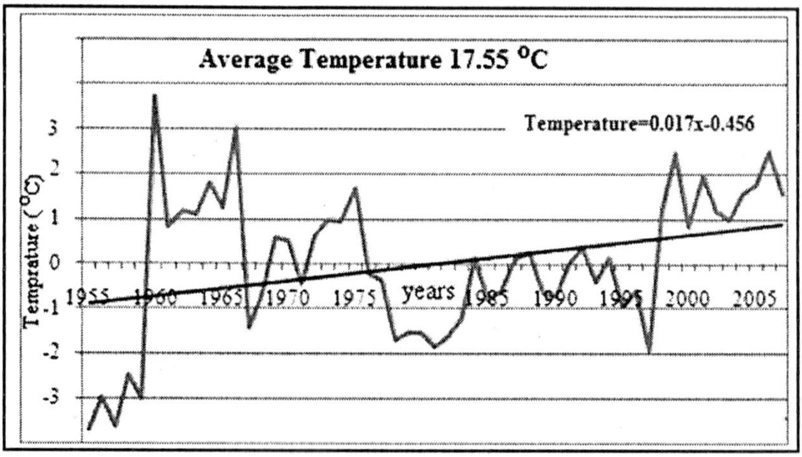

Fig. 1. Variation in average annual temperature and its trend at Almora
Source: UCCC, Interim Report, Uttarakhand Centre on Climate Change, (Kumaun University, Nainital) 2011, p.5

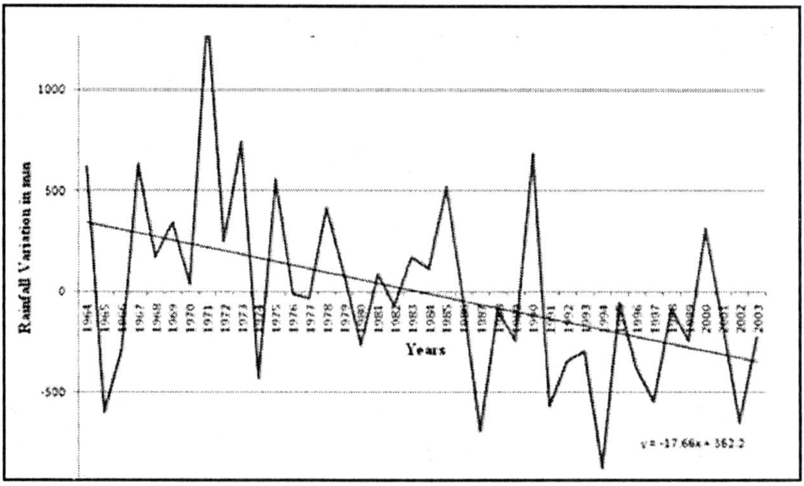

Fig. 2. Variation from average rainfall and its trend at Manora peak (Nainital)
Source: UCCC, Interim Report, Uttarakhand Centre on Climate Change, (Kumaun University, Nainital) 2011, p.6

The meteorological data of Dehradun (670m) has also indicated a change in annual average temperature of the period 1981- 2008. During 2001-2008, it has increased by 0.4°C as compared to 1991-2000 and by 0.5°C more as compared to 1981-1990. The rainfall data also show that the average annual rainfall has decreased by 149.8mm during 2001-2008 as compared to 1981-1990.Though the decrease in rainfall is continuous from 1981-1990 to 2001-2008, it is higher during the present decade as compared to 1991-2000 (Devlal, 2010).

The Interim Report on Climate also presents satellite data analysis from 1972 to 2010 of vegetation line and concludes that on an average, it has shifted by 200m towards north while in certain watersheds it has shifted by more than 320m. The report mentions that snow covered areas of the state have reduced by 18%, from 4884km^2 to 4146km^2 during 1990-99. Very significantly, the report has highlighted the fact that glacial fed mighty rivers are in the process of being converted into non-glacial rivers as the snow cover in their catchment has shrunk drastically. The report, among other impacts of climate related change, highlights the reduction in crop yield, fruit production, phonological changes in crops, trees and shrubs, shifting of fruit belts towards higher elevation and impact on pest spectrum due to climate change in the region (UCCC, 2011).

The State Action Plan on Climate Change prepared by the Forest Department and supported by United Nations Development Programme, India provides some information changing climatic conditions of the state and suggests strategies to combat these. The report presents 40 years (1968-2008) climate data of Pantnagar (244m), a place situated in terai area of the state and interestingly it underlines the fact that there is decrease in maximum temperature at the rate of 3°C/100year but an increase of 2.1°C/100year in the minimum temperature. An alarming trend has been noted in increased cloud cover duration and decrease in the rate of bright sunshine by 2hrs/100 years. Both, the increase in minimum temperature and reduction in sunshine are considered detrimental to crop production (SAP, 2011).

Retreating of glaciers is considered as a major indication of global warming. It has been reported the rate of retreat of Gangotri glacier of Uttarakhand Himalaya as more than three times in last three decades as compared to last 200 years (Liu J & Rasul G, 2007). Based on different studies, the UCCC report also highlighted the fact that important glaciers of the region have retreated in last half century. While the Gangotri glacier has reportedly retreated by 19m/year (1936-96), the Raktavarna glacier has retreated at the rate of 37m/year (1962-2004) (UCCC, 2011).

The impacts of glacial melting are not limited to the region itself but have far reaching consequences for downstream communities of north Indian plains. At the local level, the increased runoff result in flash floods during summer and lean period reduced flow will adversely impact the hydropower generation, irrigation etc., the accelerated rate of glacial melt will have serious consequences for the freshwater ecosystems of the Ganga basin, with long term impacts for biodiversity, people and livelihoods as well as regional food security (Rao, *et al.* 2012).

Occurrence of forest fires

Uttarakhand Himalaya has witnessed average temperature rise, cloud burst and long dry spells during monsoons in recent years. However one of the overlooked consequences of climate change is forest fires which have engulfed thousands of hectares of forests, leading to high loss of biodiversity and in many places, destroying indispensable source of livelihood. Warmer temperatures and drier climates make a perfect recipe for forest fires. In Uttarakhand, forest fires have been occurring during summer for the past few years. Community leaders are the opinion that a steady rise in summer temperatures which have reached unusual highs of 40°C, coupled with bad policies of forest management have leading to increased forest fires and great loss of biodiversity (Acharya A., *et al.* 2012).

An observation by local villagers also indicate that average temperatures in summer had soared to the highest in 39 years and it has lead to severe forest fires which charred 2,426 hectares of forest. In the Garhwal region, there were 665 forest fires in year 2009 (Singh V., *et al.* 2010). Another study based on the people's perception reports that more people above the age of 40 believe that forest fires have increased in last decade and also that 100% people above the age of 60 were of the opinion that forest fires have increased in the winter months (IGEP). This year (2012) more than 3000ha area has been destroyed by forest fires.

Socio-economic impacts – Agriculture, forest dependence of communities

In the fragile ecological zone in the hilly areas of Uttarakhand, human activities, including agriculture, cause extensive land degradation which, in turn, adversely affects water retention and recharge. The problem of water shortage, exacerbated by extreme weather events such as erratic rainfall, cloudbursts, unpredictable temperatures, etc. leads to detrimental effects on agriculture, forestry and even industry. Increased urbanization is putting additional pressure on limited fertile lands in the lowlands and wetlands (Sidh, S. N., Basu, S., 2011).

Women, who spend more than 50% of their time for collection of fuel wood, drinking water, and fodder for cattle, are likely to be affected more severely from alteration in climate conditions. As apart from arranging fuel, fodder and water, they produce goods and services for households that would be produced by the market in developed countries and are the breadwinners in many households, in contrast to the common perception that they are economically dependent on their male counterparts. It is therefore important that women should be put at the centre of policy making (DGIZ, 2010).

Vulnerability of the people of Uttarakhand

The people in Uttarakhand are vulnerable community as they face variety of challenges. While the state is endowed with rich forest, water and land resources, their utilization and exploitation is directly related to sustainability and fragility of resource. Low productivity, restriction of extraction and sustainability do not provide scope for extensive forest utilization and they contribute only 4% of the agricultural GDP of the state. Huge water resources are beyond the harvesting capacity of local community as enormous cost is involved in lifting it to the fields and villages from the valley bottom. Agricultural land that engages over 65% of workforce and on which 78% of the population are dependent, is largely unirrigated in the hill areas and small-fragmented holdings does not allow intensive cultivation. Moreover, this economic scenario has added restrictive conditions such as no road access to 40% villages, no LPG access to more than 80% villagers, non-connectivity and erratic electricity supply and difficulty even in the availability of the drinking water (Macchi M, *et al.*, 2011).

In addition to limited livelihood options, the people in Uttarakhand are constantly vulnerable to earthquakes (a large area under zone 4 and 5), erosion, landslides, cloud bursts, flash floods, forest fires, degradation of water resources and even drought. Communities here have nothing much to depend in case of disasters and look to outside support because even in normal times money sent by migrated household members contribute more than 50% of the requirements of the family members staying back in the village.

Coping strategies

Uttarakhand Himalaya is home to subsistence farming communities in villages and also the transhumance communities who move to higher hills in summers and come down to valleys and Terai in winters. Even the settled communities in middle Himalaya move with their livestock to temporary higher mountain summer huts. It is therefore apparent that the focus of coping strategies in Uttarakhand Himalaya has to be agriculture and animal husbandry.

Community level responses

The farming communities have perceived variations in temperature, rainfall, snowfall and incidence of crop diseases in last couple of decades. Though they may not relate it to global warming and climate changes yet their responses and coping strategies are instant as they have to depend on agriculture and animal husbandry for their livelihood on almost daily basis.

A case study of two middle Himalayan districts of Uttarakhand (Tehri Garhwal and Almora) reports that community perceives a definite change in climate by way of rainfall reduction, delay in monsoon, reduction in amount and delay in winter rainfall and snowfall, extension in duration of dry season, higher temperature, decline in frost and hailstorm and increase in crop disease and pests. The farming community is responding to these changes by adjusting their agricultural calendar on a yearly basis by delaying or advancing the sowing of rice and other crops. Winter crops such as potatoes and wheat are also planted earlier or later in some places depending on the arrival of the winter rains. Another coping strategy in district of Almora (also other places) is the revival of traditional irrigation system with water sharing rules and regulations. Another response to decreased water availability was protection of catchment areas by planting oak trees around spring catchments near villages (Lokgariwar C., 2009).

A study of two river valleys of the region brings out the relative levels of vulnerability and dictates the coping mechanisms developed in different locations. While in one valley, rural community is exploring alternative crops (e.g. Tomato in place of Apple) that will enable them to continue maintaining their current way of life, the residents of another valley, facing decline in crop production due to increased aridity in winters, are either increasingly depending on external support through remittances or seeking alternative employment or even migrating to other location (DCCPA, 2011).

A study of 6 villages of 3 districts highlights community coping strategies as mix of change in cropping pattern, switching to cash crops, revival of traditional farming methods, change in crop rotation and shift in sowing timings, income diversification and increasing dependence on government initiated employment schemes. The study also highlights the gender perspective by underlining that women who are traditionally responsible for agricultural operations are well aware of climate change and are of the opinion that local crops and varieties, and biodiversity-based agricultural inputs such as farmyard manure, enables them better to deal with weather changes. Not surprisingly, wherever women play a critical role in agricultural decision-making and natural resource management, the traditional system of farming is practiced more and this has certainly improved people's resilience in coping with the impacts of climate change (SAP, 2011). A summarized picture of perceived changes and response of the community has been presented in the Table 1.

Table 1. Community response to perceived change in Uttarakhand

Perceived change	Experienced impact on livelihood systems	Response
Erratic precipitation	Decline in agricultural productivity	Changes to agricultural calendar:delayed or early sowing and harvesting of crops, Changes in crop varieties and types Increased engagement in wage labour, Labour migration
	Crop failure	Re-sowing of crops Buying food from market Reduced meals and, Barter
Overall decreased water availability	Less flow in springs and streams; drying up of springs; lowering of groundwater levels	Re-adoption of traditional water management systems; Catchment area protection Replacing large livestock with smaller species; Less land area under cultivation; Walking longer distances to fetch water
	Reduced soil moisture Mulching	Mulching
	Decline in agricultural productivity	As above

Decreased or absent winter rains	Decreased yield from winter crops (decline in agricultural productivity)	As above
Increased frequency of intense rainfall events	Soil erosion and landslides	Watershed management; afforestation
Increase in pests and disease	Reduced production	Traditional pest management strategies (spreading cow urine, salt, or ashes; crop rotation; Setting up 'kurmula' (white grub) traps, Increased use of pesticides
Increasing temperatures	Health issues (increased incidence of vector-borne diseases)	Use of medicinal plants if available Increased reliance on western medicine
	Beneficial conditions for crops	More than one or shorter cropping cycle Introduction of new crops at higher altitudes (e.g., peanuts, mango, banana)
	Double flowering of orange and apple trees diminishing fruit quality	None
Warmer and shorter	Beneficial conditions for certain crops	As above
Winters with less snowfall	Increased incidence of pests and disease	As above
Reduction in available Workforce	Increased workload for those left behind	Less land under cultivation Move from big ruminants to smaller livestock

Source: Macchi M, Gurung A M, Hoermann B and Chaudhary D. 2011, Climate Variability and Change in the Himalayas: Community Perception and Responses, ICIMOD, Kathmandu, p.23

Policy initiatives

The State Action Plan on Climate Change has identified broad elements of the adaptation strategy in agriculture sector in Uttarakhand. These include investments in adaptation research capacity (varieties, land use systems, resource conservation technologies, pest surveillance, traditional knowledge in agriculture); changes in policies (e.g. incentives for resource conservation, integrated and diversified farming systems, promoting organic farming, pricing of resources, credit for transition to adaptation technologies); investments in infrastructure for water management and soil conservation; relocation to more productive areas and practices; greater insurance coverage for the farming operations; improved information, knowledge base, dissemination of information on climate changes and options to adapt to them; creating alternate livelihood options etc. (IGEP, 2012).

In one of the initiative, Indo-German Environment Programme has launched programme to strengthen adaptation capacities in hill community through development of climate resilient infrastructure and strengthening of alternative livelihoods, improving capacities and decision making skills of vulnerable communities on climate change adaptation, providing advisory services on disaster risk reduction, enhancing capacities of key stakeholders to assess climate change threats and opportunities, and incorporating climate adaptation measures into planning process to reduce vulnerability (Kelkar *et al.* 2008).

A study in a micro-watershed in north-western part of the region revealed the low coping capacity of people to climate variability at the individual as well as institutional level. The dependence on rainfed agriculture, non-connectivity and non- availability of formal credit, constrains the community ability to use their agricultural skills and assets more effectively. They also have limited human resources in terms of forrnal education or vocational skills, which limits their options in seeking off-farm employment opportunities. The types of responses to poor rainfall reported by households are only temporary coping measures, some of which, like selling assets or taking loans from traditional moneylenders, may actually increase their vulnerability over time by worsening impoverishment or indebtedness. However, water harvesting interventions are clearly feasible, and are being promoted by the Government of Uttarakhand, albeit in a top- down manner without always understanding the ground situation. Therefore it is necessary to develop a programme of participatory research on adaptation responses that are evolved by the affected communities themselves as such a programme can help policy- makers more effectively target resources to minimize the adverse effects of current and future water scarcity due to climate variability(Rawat *et al.*, 2011) . It has also been suggested that the climate response strategy should include key elements such as accelerating inclusive economic growth, promoting sustainable development, securing and diversifying livelihood and safeguarding ecosystem services (Rawat, 2011).

Conclusions

The available scientific evidence and people's perception clearly indicate a change in climatic conditions in Uttarakhand Himalaya. Increase in temperature, reduction in rainfall, shift in rainfall and temperature patterns, and increasing occurrence of extreme conditions are being perceived as indications of climate change in the region. While the opinion differ among the scientific community about the nature and intensity of climate change, the local community affected by the changes, have resorted to variety of coping strategies. Change of crop types and verities, change in agricultural calendar, restoration of traditional water management systems, out-migration and alternative employment avenues are some of the coping strategies being tried by the local community. There is a strong demand by the community to have accurate information about the climatic change and the possible agricultural strategies, so that they can adopt appropriate mechanism to deal with the change in climatic conditions. Thus a farmer friendly coping mechanism must be developed to support the local communities in Uttarakhand.

References

Acharya A, Vanishing Wisdom: Across Uttarakhand Traditional Systems do Disappearing Act, Tehelka Magazine 7 (5), 6 February, 2010 Retrieve 25 August, 2012, www.tehelka.com

Devlal R, Potential Threats to Uttarakhand Due to Climate Change Related Natural Disasters, in *Reflections of Climate Change Leaders from Himalaya: Natural Resource Management and Biodiversity*, edited by P DVarma (Lead India, New Delhi), 2010 pp.111-113.

Documentation of Climate Change Perception and Adaptation Practices in Uttarakhand: Northern India, Beej Bachao Andolan (Save Seed Campaign), Uttarakhand, in collaboration with Pesticide Action Network, Asian and Pacific, Penag, Malaysia, 2011 pp.21-22.

Government of Uttarakhand, State Action Plan on Climate Change-Uttarakhand: Transforming Crisis into Opportunity, first draft version, Dehradun, 2011, p.35.

Government of Uttarakhand, State Action Plan on Climate Change-Uttarakhand: Transforming Crisis into Opportunity, first draft version, Dehradun, 2011, p.71.

Hindustan Times, Global warming has no Impact on Himalayas, Hindustan Times 9 December, 2011, Retrieve 20 August, 2012, www.chimalaya.org/2011/12/09

International Centre for Integrated Mountain Development, Climate Change, www.icimod.org/climate, 2012.

Intergovernmental Panel on Climate Change, IPCC Fourth Assessment Report www.ipcc.ch/publications_and_data/ar4/syr/en, 2012.

Indo-German Environment Programme: Strengthening Adaptation Capacities in Hilly Region: Upscaling Community based Approach to Climate Change Adaptation, Almora, Uttarakhand, www.giz.de

Indo-German Environment Programme, Up-scaling Community Based Approaches to Climate Change Adaptation, Almora, Uttarakhand, www.giz.de, 2012.

Kumar K, Khanduri K & Bisht K S, Landform Systems and Physiographic Regions of Uttarakhand, in *Resources, Environment and Development of the Indian Himalaya: Issues and Strategies for Sustainable Development,* edited by, M.S.S. Rawat, D Pratap & V Singh, (Department of Geography, HNB Garhwal University, Srinagar, Uttarakhand), 2011, pp.145-162.

Kelkar U, Narula K K, Sharma V P & Chandana, U, Vulnerability and Adaptation to Climate Variability and Water Stress in Uttarakhand State, India, *Global Environmental Change* 18 (4), 2008, pp.564-574.

Liu J & Rasul G , Climate Change, the Himalayan Mountains and ICIMOD, Sustainable Mountain Development-53, International Centre for Integrated Mountain Development, Kathmandu, Winter 2007, p.11.

Lokgariwar C, Changing with the Seasons: How Himalayan Communities Cope with Climate Change (People's Science Institute, Dehradun) 2009, p.13.

Michaels P J 2010, Global Warming: Correcting the Data, *Regulation,* Fall 2008, Retrieve 16th August 2012 from www.cato.org/pubs

Macchi M, Gurung A M, Hoermann B & Choudhury D, Climate Variabilty and Change in the Himalayas: Community Perception and Responses, (International Centre for Integrated Mountain Development, Kathmandu, Nepal) 2011, pp. 21-24.

Pandey P & Vekataraman G, Climate Change Effect on Glacier Behaviour: A Case Study from Himalayas, Retrieve 20 August, 2012 , www.earthzine.org/2012/02/22.

Rawat M S S, Uttarakhand: Land and People, in *Souvenir for National Seminar on Resources, Environment and Development of Indian Himalaya: Issues and Strategies for Sustainable Development* (Department of geography, H.N.B. Garhwal University, Srinagar, Uttarakhand) 2011, pp. 46-47.

Rao P, Areendran G & Sareen R, Potential Impacts of Climate Change in the Uttarakhand Himalayas, Himalayan Documentation Centre, ICIMOD, Kathmandu, p.2 350.org, Forest Fires in Uttarakhand, India, Retrieve 25 August 2012, www./350.org

Rawat R B S, Impact of Climate Change on Livelihood Options in Himalayan Region: Issues, Challenge and Strategies, in *Resources, Environment and Development of the Indian Himalaya: Issues and Strategies for Sustainable Development,* edited by, M.S.S. Rawat, D Pratap & V Singh, (Department of Geography, HNB Garhwal University, Srinagar, Uttarakhand), 2011 pp.143-144.

Singh V, Forest Fires and Climate Change: A Case Study from Kumaun Uttarakhand Himalaya, in *Reflections of Climate Change Leaders from Himalaya: Natural Resource Management and Biodiversity*, edited by P D Varma (Lead India, New Delhi), 2010, p.104.

Sidh S N & Basu S, Women's Contribution to Household Food and Economic Security: A Study in the Garhwal Himalayas, India. *Mountain Research and Development* 31(2), (2011) 102-111.

State Action Plan on Climate Change, Uttarakhand: Transforming crisis into opportunity, First draft version, Government of Uttarakhand, 2011, p.31.

Shukla P R, Sharma, S K, Ramana, P V & Bhattacharya, S, Introduction, in *Climate Change and India: Issues, Concerns and Opportunities,* edited by P R Shukla, S K Sharma & P V Ramana (Tata McGraw-Hill Pub Co Ltd, New Delhi) 2002, p.1.

Singh S P, Climate Change in Relation to the Himalayas, Retrieve 1 July, 2012. from www.climate- leaders.org/wp-content/uploads/climatechange-spsingh.pdf

United Nations Environment Programme, Climate Change, www.unep.org, 2012.

UNEP, Recent trends in melting glaciers, tropospheric temperatures over the Himalayas and summer monsoon rainfall over India, Division of Early Warning and Assessment (DEWA), United Nations Environment Programme (UNEP), Nairobi, Kenya, 2009, p.2.

UCCC, Interim Report, Uttarakhand Centre on Climate Change, (Kumaun University, Nainital) 2011, pp. 5-6.

UCCC, Interim Report, Uttarakhand Centre on Climate Change, (Kumaun University, Nainital) 2011, pp. 9-26.

UCCC, Interim Report, Uttarakhand Centre on Climate Change, (Kumaun University, Nainital) 2011, pp.9-26.

Uttarakhand: Diagnostic Study of Building a Mountain State 2000-2010, Deutsche Gesellschaft fur Internationale Zusammenarbeit (GIZ) GmbH and Doon University, Dehradun, pp.25-26.

Vegetation and taxonomic profile of the flora of Indian cold desert: A highly fragile and vulnerable mountain ecosystem in the country

Achuta Nand Shukla[1] *& S.K. Srivastava*

Botanical Survey of India, Northern Regional Centre,
Dehradun-248 195, Uttarakhand, India
[1][E-mail- achutbsi@gmail.com]

Introduction

In India, the Trans-Himalayan zone lies in the rain shadow of the main Himalayan range is usually described as 'High Altitude Cold Desert'. It is more extensive outside the Indian Territory adjoining Tibet and in India it covers only about 2% of total land surface[1]. Within Indian limits bulk of the cold desert lies in Ladakh region, in Jammu & Kashmir, Lahaul & Spiti, and Kinnaur (small northern portion of Kalpa and Baspa valleys) in Himachal Pradesh and small portions of Uttarkashi (Nelang valley) and Chamoli (Mana and beyond and Niti beyond Malari) districts in Uttarakhand i.e. northern tip of Garhwal Himalaya. The total area of cold desert in India is approximately 98,660 sq. km i.e. about 82,665 sq. km in J. & K., about 15,000 sq. km in Himachal Pradesh and about 1000 sq. km in Uttarakhand (Murti, 2001). Geologically the Himalayas fall into 3 series, northern, central and southern. The soils of Ladakh have been classified as grey, light arid, sandy or sandy loam of low fertility with appreciable quantities of clay at some places (Bhat, 1965). Extending over the north of Himalayan ranges, Trans-Himalayan zone lies in rain shadow of main Himalayan range. The cold deserts are characterized by extremely low temperature -45°C, low rainfall ranging from 500-600 mm annually, forms a plateau extends from 4500-6000 m altitude in western Himalayas dominated by annual and perennial herbs, few stunted shrubs and bushes (Chandra Sekar and Srivastava, 2009). The climate is characterized by great extremes of heat and cold, coupled with excessive dryness. Absence of humidity helps to keep the atmosphere clear. In the summer, the heat during day time is so intense and it is so cold at night that the mountains crack and crumble (Kachroo *et al.*, 1977).

Fragility of mountain ecosystems and their vulnerability to adverse effects of climate are well established. Increase in human pressure and activities pose serious threat to the survival of the floristic diversity and ecosystem as well. This human pressure and over exploitation of natural resources have made the trans-

Himalayan cold desert, one of the world's most fragile and threatened ecosystems (Murti, 2001).

Materials and methods

Intensive plant explorations, and in-depth surveys of the area were made from 2011 to 2012. Specimens were collected from different localities of the cold desert *viz.*, Rohtang, Tsomoriri (Fig. 1) Baralacha La, Taglang La, Nakila, Salsal, Leh, Hemis, Changthang, Rangdum, Pezila, Tsokar Lake (Fig. 2) Zanskar, Bothkharboo, Chattru, Chotadara, Lossar, Kaza, Pin Valley, Kibber Wildlife Sanctuary, etc. The specimens have been deposited in the herbarium of Botanical Survey of India, Dehradun (BSD).

Fig. 1. Tsomoriri Lake-World's Highest Ramsar Sites

Fig. 2. Tsokar Lake, Ladakh

Results and discussion

Vegetation

The flora of cold deserts belts come under alpine and high alpine zones and are dominated by annual and perennial herbs, followed by few stunted shrubs and bushes. Trans Himalayan cold desert has a characteristic vegetation differing from other parts of Himalayas due to prevailing unique conditions and physiographic (Nair, 1964). Vegetative growth starts at the commencement of summer when the melting snow provides abundant moisture. Due to brief growing period, majority of plant species are in full bloom during July-August but start disappearing by the end of September (Murti, 2001). Alpine mountain slopes, meadows and alpine pasture lands abound in varieties of alpine and high alpine flowers making trans Himalayan barren mountains lush green for brief summer period. Trans-Himalayan region support very scanty and highly specialized vegetation characteristic of cold desert (Fig. 3). The vegetation of cold desert broadly categorized into four groups (Murti, 2001).

Fig. 3. A view of Typical Cold Desert

Temperate vegetation

This vegetation is very much localised and is available between tract of Thirot and Udaipur of Lahaul valley. The vegetation is characterized by woody elements (Aswal, and Mehrotra, 1979). Main composition of this vegetation in woody groups are *Pinus wallichiana*, *Picea smithiana*, *Cedrus deodara*, *Betula utilis*, *Juniperus macropoda*, *Juniperus recurva*, *Juglans regia*, *Prunus armeniaca* and different species *Salix, Popolus, Malus, Sorbus,* etc.

Major shrub species of this zone are *Berberis pachyacantha*, *Cotoneaster* sp., *Ephedra gerardiana, Fraxinus xanthoxyloides, Hippophae rhamnoides var. turkestanica, Hippophae salicifolia, Juniperus communis, Lonicera sp., Rosa macrophylla, Rosa webbiana,* etc. form the shrubby growth. *Myricaria germanica* frequently grows along riverside. The herbaceous elements include *Anemone* spp., *Aquilegia fragrans*, *Aquilegia pubiflora*, *Codonopsis ovata*, *Galium aparine*, *Geranium pratense*, *Inula grandiflora*, *Podophyllum hexandrum*, *Primula denticulata*, *Ranunculus laetus*, *Rheum spiciforme* (Fig. 4), *Oxyria digyna*, *Rumex nepalensis*, *Achillea millifolium*, *Capsella bursa-pastoris*, etc. Some other important herbaceous elements of meadows and slopes are *Anemone rivularis*, *Brachyactis roylei*, *Cirsium wallichii*, *Heracleum lanatum*, *Thymus linearis*, *Phlomis bracteosa*, *Origanum vulgare* and species of *Anaphalis*, *Arenaria*, *Astragalus*, *Chenopodium*, *Geranium*, *Impatiens*, *Lepidium*, *Medicago*, *Melilotus*, *Pedicularis*, *Silene*, *Trifolium*, *Trigonella*, *Senecio* etc. The common herbaceous elements in marshy places and along streams include *Deschampsia caespitosa*, *Halerpestis sarmentosa*, *Danthonia cachemyriana*, *Elaeocharis palustris*, *Polygonum hydropiper*, *Ranunculus natans*, *Carex* spp., *Juncus* spp., *Veronica* spp., *Kobresia* spp., *Astragalus melanostachys*, *Caltha palustris*, *Parnassia laxmanii*, *Sagina saginoides*, *Epilobium* spp., etc. Occasionally some tufted plants, typical of cold desert areas are also found. Some important ones are *Arenaria festucoides*, *Cerastium cerastioids*, *Dianthus angulatus*, *Minuartia kashmirica*, *Rosularia alpestris*, *Silene stewartii*, etc.

Fig. 4. Rheum spiciforme Royle

Alpine mesophytes

This zone is also characterized by high humidity and more rainfall. Flora of Suru valley (Fig. 5) and parts of Lahaul valley shows the characteristic of alpine mesophytes (Aswal and Mehrotra, 1981). Majority of plants growing in this zone, are also found in temperate regions. Plants restricted to moist regions, in depressions and slopes which are moistened by melting snow and glacial streams. The most conspicuous character of the vegetation is the cushion like habit of plants. It protects these plants from the intensely cold dry winds. The common mesophytic plant species are *Podophyllum hexandrum*, *Lavetera kashmiriana*, *Lotus corniculatus*, *Astragalus rhizanthus*, *Verbascum thapsus*, *Oxyria digyna*, *Capsella bursa-pastoris*, etc. Only about 10% of alpine mesophytes are typical cold desert plants like *Delphinium brunonianum*, *Potentilla* spp., *Leontopodium* spp., *Taraxacum* spp., etc. which are widely distributed in high altitude areas.

Fig. 5. Vegetation along Suru Valley, Kargil

Along river courses and in marshy depressions *Salix daphnoides*, *Myricaria elegans* are found. Along the sandy banks of the Drass river and other streams there is a rich growth (Joshi, 1952) of *Phragmites karka* associated with *Equisetum* spp., and *Sonchus oleraceous*. The slopes are covered with plants like *Bistorta affine*, *Potentilla bifurca*, *Pedicularis hookeriana*, *Parnassia laxmanni*, *Geranium pratense*, *Brachyactis roylei*, *Brachyactis pubescens*, *Anaphalis nepalensis*, *Galium boreale*, *Euphrasia vulgaris*, *Lactuca tatarica*, etc.

Desert vegetation

This zone is characterized by little rainfall, low humidity, extreme fluctuation of diurnal temperature and high velocity winds etc. Long stretches of Ladakh and Lahaul-Spiti regions *viz.*, Central Ladakh, Puga areas, Tangtse, Kaza, Pin valley are characterized by stunted, cushion type desert vegetation of tufted plants (Murti, 2001). Due to high velocity winds, which are constantly blowing in these areas, the plants tend to become prostrate, thick woolly, cushion forming, bushy, hardy, spongy and with long deep rooted roots and small leaves (Naqshi *et al*., 1989). The herbaceous species show a mixture of Tibetan, Siberian, Chinese and Afghanistan flora. Majority of plant species are characterized by stunted, cushioned, prostrate, thick woody, spongy and long roots and succulent small leaves. Common plant species which are growing around barren valleys of Indus, Changthang (Fig. 6) and Spiti, are *Polygonum avicule*, *Atriplex crassifolia*, *Corydalis flabellata*, *Caragana pygmea*, *Echinops cornigerus*, *Peganum harmala*, *Capparis spinosa*, *Berberis ulicina*, *Tanacetum* sp., etc. Similarly desertic elements found growing around high passes like Khardungla (5520 m), Changla (5286 m), Penzila (4204 m), Taglangla (5255 m) and Kunzum pass (4500 m) are *Thylacospermum caespitosum*, *Acantholimon lycopodioides*, *Arnebia euchroma*, *Caragana pygmea*, *Dracocephallum heterophyllum*, *Meconopsis aculeata*, *Primula macrophylla*, *Rhodiola imbricata*, *Rosularia alpestris*, *Waldhemia tomentosa*, *Euphorbia tibetica*, *Arenaria bryophylla*, *Nepeta floccosa*, *Christolea crassifolia*, *Saussurea bracteata*, *Saxifraga* spp. On rocky and gravelly sites the common plants are *Ephedra gerardiana*, *Caragana pygmaea*, *Rosa webbiana*, *Physochlaina praealta* and *Stachys palustris*. In areas with alkaline swamps and borax deposits, *Triglochin palustris*, *Triglochin maritima*, *Carex alpina*, *Polygonum sibiricum* etc. are commonly growing. In a few cultivated sites some common weeds are *Chenopodium album*, *Melilotus alba*, *Ranunculus sceleratus*, *Ranunculus pulchellus*, *Convolvulus arvensis* etc. Along streams *Myricaria elegans*, *Hippophae rhamnoides*, *Rosa webbiana*, *Salix fragilis*, *Salix alba*, *Potentilla anserine*, *Oxytropis macrophylla*, *Lancea tibetica*, *Pedicularis longiflora* etc. form conspicuous association. The upper portion of the Indus valley is full of silt and almost without any vegetation.

Fig. 6. Changthang area of Ladakh

Although there are no natural forest in Ladakh and parts of Lahaul-Spiti, there are frequently thickets along streams and along river belts. These have a typical association of *Myricaria* sp., *Hippophae* sp., *Rosa* sp., *Lonicera* sp., etc. The introduced forest species are *Populus, Salix, Prunus, Malus*, etc.

Typical desertic elements are found in the areas beyond the transition zone. These areas receive none or very scanty rainfall with extreme fluctuation of diurnal temperature. Due to high velocity winds, constantly blowing in these high altitude areas, the plants tend to become prostrate, thick woolly, cushion forming, bushy, hardy, spiny and with deep penetrating long roots and small leaves. Some common typical desert species are *Thylacospermum caespitosum, Acantholimon lycopodioides, Echinops cornigerus, Lindelofia anchusoides, Tanacetum tibeticum, Nepeta floccose, Arnebia guttata, Lancea tibetica, Lepidium apetalum, Elaeagnus angustifolia, Potentilla anserine, Sedum ewersii, Saxifraga sibirica, Waldheimia tomentosa, Dianthus anatolicus, Oxytropis lapponica, Potentilla multifida, Plantago minima, Sedum tibeticum, Arabis tibetica, Corydalis crassifolia*, etc.

Oasitic vegetation

Plants of this zone are generally found near habitations like Kargil, Leh, Khoksar, Keylong, Kaza, Trilokinath, Jispa, etc. and are cosmopolitan in nature. This type of vegetation is represented by a variety of exotic as well as indigenous species, growing near habitation, along water channels, streams and in moist places. The common elements are *Lancea tibetica, Pedicularis longiflora, Gentiana algida, Arabidopsis himalaica, Ranunculus pulchellus, Geranium collinum, Geranium sibiricum, Medicago lupulina, Astragalus densiflorus, Rubus saxatilis, Lomatogonium carinthiacum, Chenopodium foliosum, Primula sibirica, Mentha longifolia, Lycium ruthenicum, Cuscuta europea, Epilobium roseum, Vaccaria pyramidata, Trifolium pratense, Sophora alopecuroides, Melilotus alba, Melilotus officinalis, Potentilla ambigua, Potentilla bifurca, Elsholtzia densa, Nepeta tibetica*, etc.

Trees, with the exception of Hippophae, are practically all introduced. The common species are *Salix, Populus, Juglans, Pyrus* and *Prunus*. Small groves of *Juniperus recurva* are sometimes found here and there.

Adaptation for survival strategies

The plants of cold deserts exhibit a number of ecological, morphological and physiological adaptations which help them to counteract the impact of harsh climate prevailing in these regions (Murti, 2001). These plants are capable of establishing themselves in cold arid regions as far as there is dry soil or substratum available to provide them anchorage and also remains free from ice or snow just for few weeks in a year. Some peculiar adaptation techniques of these plants are discussed below.

Cushion-forming habit

Cushion, clump or mat-forming habit is very common in cold desert plants. Such plants are perennial, short and sturdy with woody stem and deep root system capable of penetrating rock crevices and fissures to provide firm anchorage and nutrition to the plant. Such a habit protects from the strong wind action and its drying effects, strong thermal radiations, loss of water through transpiration, in maintaining the balance of temperature fluctuations between air and soil and from the continuous pressure of snow layer which may be several feet thick for months together e.g. *Acantholimon lycopodioides, Thylacospermum caespitosum, Arenaria bryophylla,* and species of *Astragalus, Androsace, Draba, Sedum, Saxifraga* etc.

Diminutive or miniature habit

Although the cold desert plants are generally dwarf and stunted, some of them are so significantly reduced, that one may not even notice them in fields. Species like, *Pleurogyne brachyanthera, Gentiana thomsonii,* and *Taraxacum bicolor* are often barely 1-2 cm tall with a solitary flower. *Ranunculus tricuspis, Anemone imbricata, Lancea tibetica, Saxifraga parva,* are some other small sized species. Few *Saussurea* species are an example of diminutive plants of cold desert. In case of *Astragalus heydei, Corydalis crassissima, Thermopsis inflata,* and *Dracocephalum heterophyllum* develop deep penetrating permanent root-stocks from which annual branches are produced down and they bear leaves and flowers in clusters just above the stones or rocks.

Bushy habit

The number of woody tall plants in cold arid region is exceedingly low. *Caragana pygmaea, Ephedra gerardiana, Hippophae rhamnoides, Myricaria prostrata* and *Lonicera hispida* form dense bushy habit with woody branches barely attaining 30-150 cm.

Protective covering of hairs

These hairs are very dense and form a felt like coating on the entire surface of the exposed plant and acts as a thermal blanket. Besides providing thermal security, these hairs impart a silver-grey or white appearance to the plant which helps in reflecting the solar radiations reducing thereby the harsh impact of sun's rays. *Astragalus munroi* (Fig. 7), *Saussurea gossypiphora* and *Soroseris glomerata* having profuse wool coating on the vegetative and floral parts.

Floristic diversity

Kachroo *et al.* (1977) enumerated 611 species from Ladakh. Aswal & Mehrotra (1994) recorded 985 species from the Lahaul-Spiti in Himachal Pradesh (Aitchison, *et al.*, 1868, Singh and Gohil, 1972, Aswal, and Mehrotra,1980, Srivastava, *et al.*, 1981). Based on recent collections, herbarium records and reports published, the flora of the cold desert comprises *ca* 1405 species of flowering plants, distributed in 490 genera and 98 families, including 8 species of gymnosperms (Balapure, 1982, Misri, 1982, Aswal and Mehrotra, 1985, Aswal and Mehrotra, 1994).

Fig. 7. *Astragalus munori* Bunge

Endemic species

Ferula jaeschkeana, Lanceatibetica (Fig. 8), *Poa ladakhensis, Potentilla thomsonii, Ranunculus trivedii, Saussurea graminifolia, Silene stewartii, Thalictrum rutaefolium, Waldheimia stoliczkei, Allium loratum, Anaphalis royleana, Arenaria stracheyi, Astragalus munroi, Astragalus zanskarensis, Berberis ulcina, Christolea stewartii, Corydalis crassifolia, Dactylorhiza hatagirea, Draba cashmirica.*

Fig. 8. *Lancea tibetica* Hook.f. & Thomson

Threatened species in cold desert

Allium carolinianum, Astragalus malacophyllus, Cremanthodium decaisnei, Draba cachemirica, Eremurus himalaicus, Galium tibeticum, Geranium pseudo-aconitifolium, Oxytropis sericopetala, Poa ladakhensis, Primula obtusiloba, Rhodiola brunonii, Rhodiola tibetica, Saussurea bracteata, Stellaria pusilla, Thermopsis inflata (Fig. 9), *Thlaspi cochlearioides, Thylacospermumca-espitosum* (Fig. 10), *Viola kunawarensis,* etc.

Fig. 9. *Thermopsis inflata* Cambess.

Fig. 10. *Thylacospermum caespitosum* (Cambess.) Schischk.

Medicinal and economically potential species

Cold desert has rich wealth of plants having economic potential as crop, medicinal, ornamental and ethnobotanical plants. Number of plant species found in the area is being used in curing various ailments by the local people also known as 'Amchis system' of medicine and many species are used for various other purposes (Chaurasia *et al.*, 2007).

Aconitum heterophyllum, Aconitum violaceum, Artemisia maritima, Arabis tibetica, Astragalus candolleanus, Berberis ulcina, Capparis spinosa, Carum carvi, Chenopodium botrys, Convolvulus arvensis, Dactylorhiza hatagirea, Dracocephalum heterophyllum, Ephedra gerardiana, Ferula jaeschkeana, Gentiana kurroo, Hippophae rhamnoides, Hyoscyamus niger, Meconopsis aculeata, Papaver nudicaule, Picrorhiza korrooa, Podophyllum hexandrum, Rheum emodi.

Economic plants

The wild edible fruits of *Capparis spinosa, Cotoneaster falconeri, Ephedra gerardiana, Hippophae rhamnoides, Ribes orientale* and *Rosa hookeriana* are consumed by local people. Plants or their parts are cooked as vegetables from *Allium carolinianum, A. stoliezki, Chaerophyllum acuminatum, Chenopodium album, Cicer microphyllum, Eremyrus himalaicus, Mentha longifolia* and *Rumex acetosa.*

Astragalus chlorostachys, A. peduncularis, Caragana versicolor, Cicer microphyllum, Hippophae rhamaoides ssp. Turkestanica, Medicage falcata, Myricaria albiflora, Myrtama elegans, Populus ciliata, Salix denticulata, S. fruticulosa and *Trigonella emodi* are used as fodder and fuel purposes. The wood of *Juniperus recurva, Populus ciliata* and *Salix denticulata* are used for timber and implements. *Arnebia euchroma, Geranium pratense* and *Onosmo hispidum* are used as dye. The incense and perfumes are obtained from *Chrysanthemum pyrethroides, Delphinium brunonianum* and *Waldheimia glabra.* Bushes of *Hippophae rhamnoides* are used for fencing.

Acknowledgement

Authors are thankful to the Director, Botanical Survey of India, Kolkata for the encouragement and facilities.

References

Aswal B S and Mehrotra B N, New records of plants from Himachal Pradesh, India, *Indian J. Forest.* 2(4): (1979) 322.

Aswal B S and Mehrotra B N, Contribution to the flora of Lahaul Valley (North-West Himalaya)-III. A note on the nomenclature of plants, *J. Econ. Taxon. Bot.* 2: (1981) 236.

Aitchison J E T, Lahaul, its flora and vegetable products etc. *J. Linn. Soc. Bot.* 10: (1868) 69-101.

Aswal B S and B N Mehrotra, Contribution to the flora of Lahaul Valley (North-West Himalaya)-I. Some new plant records, *Indian J. Forest.* 3(2): (1980) 154-155.

Aswal B S and B N Mehrotra, Contribution to the flora of Lahaul Valley (North-West Himalaya)-V. Phytogeographical aspects, *J. Econ. Taxon. Bot.* 7(2): (1985) 299-307.

Aswal B S and B N Mehrotra, *Flora of Lahaul-Spiti.* Bishen Singh Mahendra Pal Singh, Dehradun (1994).

Bhat G M, The soils of Kashmir. Bull. Agri. Deptt. J. and K. Govt. Srinagar. (1965).

Balapure K M, Some new records and additional localities of plants from Ladakh, *J. Econ. Taxon. Bot.* 3(1): (1982) 187-191.

Chandra Sekar K and S K Srivastava, *Flora of Pin Valley National Park, Himachal Pradesh.* Botanical Survey of India, Kolkata (2009).

Chaurasia O P, Ahmed Z and Ballabh B, Ethnobotany and Plants of Trans-Himalaya. Satish Serial Publishing House, Delhi (2007).

Joshi A C, Aquatic vegetation of Lahaul. *The Palaeobotanist* 1: (1952) 277-280.

Kachroo P, B L Sapru and U Dhar, *Flora of Ladakh.* Bishen Singh Mahendra Pal Singh. Dehradun (1977).

Murti S K, Flora of Cold Desert of Western Himalaya 1. Monocotyledons. Botanical Survey of India. Calcutta (2001).

Misri B, A preliminary survey of grasses and legumes of Ladakh, *Austr. Plant Introduction Review.* 30: (1982) 37-45.

Nair N C, Some plant records for Lahaul-Spiti, *Bull. Bot. Surv. India* 6: (1964) 151-153.

Naqshi A R, N Y Malla and G H Dar, Plants of Ladakh, Nubra, *J. Econ. Taxon. Bot.* 13(3): (1989) 539-560.

Singh G and R N Gohil, Some new records to the flora of Ladakh, *J. Bombay Nat. Hist. Soc.* 73(3): (1972) 487-490.

Srivastava T N, N C Shah, D P Badola and O P Gupta, New records of flowering plants of Ladakh. *Indian J. Forest.* 4(2): (1981) 138-141.

A threat to agro biodiversity and its conservation in Uttarakhand Himalaya due to commercialization of agriculture

*Mohammad Rais & Nehal A. Farooquee**

CSIR-National Institute of Science, Technology and Development Studies
K.S. Krishnan Marg, New Delhi – 110012, India
* School of Extension and Development Studies
Indira Gandhi National Open University, Maidan Garhi, New Delhi – 110068, India
[E-mail: mohammad_rais@hotmail.com, *nafarooquee@ignou.ac.in]

Introduction

A key area of concern in the Himalayan agricultural systems is threat to agro biodiversity and conservation practices from commercialization of agriculture. In Asia, the Green Revolution has had an almost pervasive impact, and India is no exception to this (Evenson & Gollin, 2003); but in the Himalayan region, farmers have enjoyed little benefits from formal crop breeding programmes, this region in particular has been described as achieving a very partial improvement in agricultural production unlike the plain areas of Uttarakhand, and the yield increase per unit area through introduction of high yielding varieties has been especially very limited in this region (Zardhari, 2000; Farooquee and Maikhuri, 2009). The limited contribution of agricultural research (including crop improvement) has also been blamed on lack of involvement by small farmers (Farnworth *et al.*, 2003). Farmers are excluded because of a linear approach in which research is seen as a source of technology, extension as the delivery mechanism, followed by a straightforward adoption by farmers (Bruin & Meerman, 2001; Röling, 1996). In order to understand this changing scenario of loss of agro biodiversity and traditional conservation practices, an empirical study was undertaken to understand the cause and consequences of declining the traditional agro-biodiversity and implications in Uttarakhand Himalaya in India.

Uttarakhand is an Indian mountain state, spread over 53,485 square kilometers has a population of about 8.4 million (2001 census), out of which 74% live in 15,024 villages. Agriculture is the mainstay of this rural population. The formidable obstacles to agricultural development include a scarcity of arable land and scanty irrigation. The government owns 66 per cent of the land which is under forest. The net sown area (excluding Nagar and Hardwar districts, which are plains areas) is about 13 percent of the total reported area. This mountain region has diverse agro-climatic features, including a flat area on the foothills called *tarai*, mid mountain and high land nearing the snow line. Though the mountainous part of the state is food deficient, but there is a good repository of diversified agricultural traditions and products. There is greater awareness now for management of natural resources worldwide and specific areas are protected in the form of sanctuary, national parks and biosphere reserves particularly for achieving the goal of in-situ conservation of biological diversity. In this endeavor there are increasing interests for conservation and management of traditional crop diversity in natural condition (Nautiyal and Kaechele, 2007).

On-farm conservation is "the conservation and management of cultivated species in the surrounding environment where they have developed their distinctive properties" (Maxted *et al.*, 1997). On-farm conservation is considered to be the sustainable management of genetic diversity of locally developed traditional agricultural, horticultural and agrisilvicultural cultivation systems. In the Himalayan highlands of India, farmers predominantly practice traditional agriculture where it meets their needs. Local communities also possess a great deal of ethno-botanical knowledge related to traditional crops. The farmers of Uttarakhand have also conserved the plant genetic resources and the methods of cultivation in the face of massive erosion of crop diversity all over India. The genepool of traditional crops has also been continuously enriched by geneflow across the Himalayan states and countries through trading of grain and exchange of seeds (Farooquee and Maikhuri, 2009). However, in the recent past, lot of erosion in the landrace diversity has been observed mainly due to habitat degradation and shift to modern high yielding varieties (HYVs). In some areas the traditional crop landraces are facing danger of complete extinction and consequently the ecological and economic security of the traditional farming systems of this region appears to be in jeopardy (Maikhuri *et al.*, 1997; Nautiyal *et al.*, 2002).

The approach in this manuscript raises research questions and conceptual concerns posed by different academic disciplines including agronomy, economics, ecology, anthropology and women's studies. It tries to avoid the tendency to compartmentalize disciplines and methodologies by pointing out the various types of roles the food crops and cash crops play in cultural and biological systems. It also tries to bridge the gap between class and gender perspectives by illuminating feminist strategies in people's designs for livelihood. One of the objectives of this paper is primarily to understand other issues around the food security, conservation and commercialization of agriculture. The general picture that emerges in the mind of the readers, and which forms perspective of this manuscript, is that agriculture is determined by the diversity and dynamism of local food systems embedded in specific cultures and ecosystems. Developing food policies, farming practices and community relations compatible with this understanding of diversity are critical not only to the conservation of the uncultivated biodiversity of Uttarakhand but also for the formulation of conservation policies at large. Based on a series of field visits, practical experiments and social surveys this paper tries to understand the meaning of diversity and conservation in the life of small farmers, some facts about uncultivated foods and their perceptions about cash crops and commercialization of agriculture.

Materials and methods

The study area is located in mid high altitude of Central Himalaya (30°32′18″N and 79°28′02″E) (Figure 1) was selected for the present study as a representative site of traditional landrace-based cultivation in the region. Altitudinal limits ranged between 1500 and 2000 m asl. The year consists of three distinct seasons: summer (April – June), rainy (July –September) and winter (October- March). Average rainfall is about 925mm, of which about 51% is received in a short period of two months (July – August) featuring a strong monsoon influence. Monthly maximum and minimum temperatures range between 28°C to 16°C, and 9.5°C to 5.0°C, respectively. Average land holding size is <1.0ha per household. Besides traditional agricultural crops, a variety of fruit trees (apple, apricot and walnut etc.) are planted on the margins of the rainfed terraces. Seasonal and off-season vegetables such as, cucurbits, ginger, cabbage, green vegetables, etc., are cultivated in the kitchen gardens. The average area under kitchen garden is about 0.012 ha/household. Agro ecosystems and cropping patterns in the rainfed agriculture on steep terraces is the predominant form of land use. All the households were involved in agriculture. The rainfed agriculture in all the villages is practiced on almost two equal halves of agricultural land locally called as 'sari' with different crop compositions. To maintain soil fertility, the tradition is to keep one half of the land fallow during winter season alternately over a period of two years. Therefore three major crops can be harvested biannually from rainfed agriculture. In irrigated land, two crops, a summer crop and a winter crop, were harvested in a year. The crop rotations, cropping patterns and crop composition, practiced in the region have also been studied.

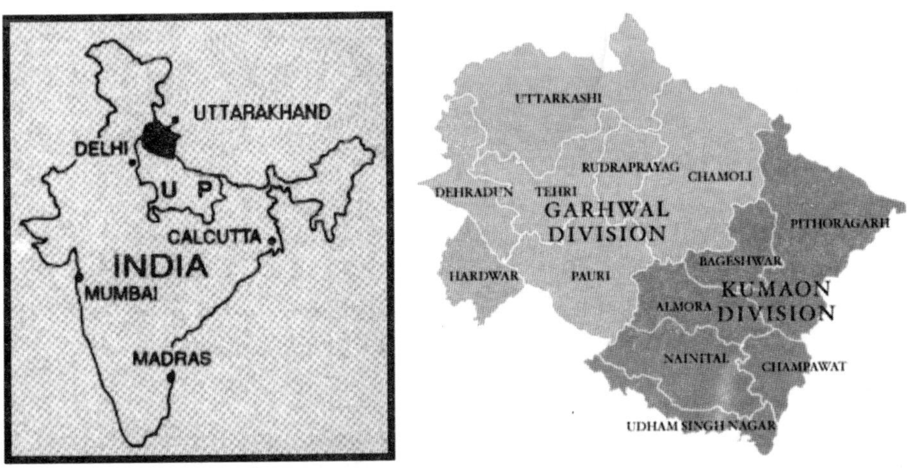

Fig. 1. Location Map of Uttarakhand

About 80 percent of people of the Central Himalaya practice subsistence agriculture. Land holdings are small and fragmented; *per capita* cultivated land is

0.2 ha. Terraced slopes, covering 85 per cent of total agricultural land are largely rainfed, while the remaining 15 per cent area in the valleys is irrigated. Mixed cropping is common under rainfed conditions. The cropping patterns are built around two major cropping seasons, *viz., Kharif* (April-October) and *Rabi* (October-April) generally up to 1800 m amsl and, up to 2000 m amsl in some places. At higher altitudes (>2000 m amsl), only summer season crops (April-October) are raised. Over 40 crop species, often erroneously termed lesser known crops and comprising of cereals, millets, pseudo cereals, pulses, oil seeds, tubers, bulbs and spices are cultivated.

The study area for fieldwork was done in Kumaun and Garhwal, the two geographical and cultural regions of Uttarakhand. In Kumaun hills it was the Tarikhet block of Almora district and in Garhwal hills, Ukhimath block of Rudraprayag district. Using defined criteria, 349 households in 28 villages of Tarikhet block of the Almora district and 261 households in 15 villages of Ukhimath block were surveyed during September 2003 to December 2005. All the households were surveyed to determine average land holding size, area under different crops, crop compositions, cropping patterns and crop rotations. The information was collected through household survey using structured questionnaire, informal discussions with knowledgeable members of the households. To gather the above information, each household was visited at least 2-3 times during the study period. It is very difficult to measure diversity of agricultural crops and ranking them in order of rarity, endangered or extinct on one hand and prejudged the farmers' behaviours and their thinking on the other. But we can measure and investigate farmer behaviour through long term observations, inquiries and socioeconomic survey methods. The data have been used to analyze the cash income generated and energy status (surplus or deficit) associated with (i) traditional coarse grains, (ii) local Landraces, and (iii) high-yielding crops in the region. The analysis was carried out using standard agro-biodiversity indicators such as Shannon diversity index.

Results and discussions

Cropping diversity

The state of Uttarakhand produces about 40 different crop species of cereals, pseudo cereals; millets, pulses, oilseeds etc. and many varieties are cultivated throughout the latitudinal gradient (Semwal *et al.*, 2001; Rais and VanLoon, 2006). So much so, that in the monsoon, typically farmers of the lower and middle Himalayas can harvest twelve crops, popularly known as *Bara Naza* (twelve grains) from tiny holdings. Mostly, in an average field, three or four subsidiary crops are raised with one main crop. It is still very common to see farmers cultivate mandua (Finger millet), Gahat *(Horesgram)* Urad (*Black gram)*, Soyabeans, Riains, (*Adjuki bean*) and Sonta *(Cowpea*) in the same field. Kodo occupies the middle portion of the field and pulses are sown in the upper and lower portions, where soil is less fertile and prone to soil erosion. This practice is used traditionally to utilize the difference in the fertility of the soil for economic advantage and simultaneously conserve the soil. The traditional crops which are not so popular are: Chaulai (*Amaranth*), Ogal and Phapar (Buckwheat) and Cheena (Hog-millet). Besides being rich in protein content, have good productivity on inferior soil and are more capable to withstand the vagaries of nature. For instance, the average productivity of Chauli is around 15 quintals per hectare, Ogal 13.5 quintal per hectare and Phapar 12.5 quintals per hectare.

Over the generations, farmers in Uttarakhand have been cultivating traditional cultivars of a number of cereals, pulses, vegetables and fodder trees for their livelihoods (Tables 1, 2 & 3). Traditionally the small farmers only cultivated the traditional crops in small quantities, in spite of their urgent need to grow more cash oriented crops to meet the growing requirement of their families. Farmers in Uttarakhand emphasized the role played by their traditional crops in ensuring household food security and maintaining the cultural heritage of their tradition (Farooquee and Maikhuri, 2009). Out of the 67 predominant food crop species of the Uttarakhand Himalaya about 34 species (comprising of cereals, pseudo cereals, millets, pulses, oil yielding crop and different kinds of vegetables) are grown in the traditional agro ecosystems of the Valleys. The traditional crops of the valley are *Oryza sativa, Triticum aestivum, Hordeum vulgare, Eleusine coracana, Echinochloa frumentacea, Setaria italica, Panicum miliaceum, Amaranthus* spp., *Fagopyrum esculentum, F. tataricum, Phaseolus vulgaris, Vigna mungo, Macrotyloma uniflorum, Glysine max* (local black seeded types), *Brasscia compestris, Perilla frutescens, Sesamum indicum*, many local vegetables (*Cucurbits, Trigonella* spp., *Beta* spp.) etc. Besides the improved types of some other crops viz., *Pisum sativum, Solanum tuberosum, Zea mays* etc. are also grown.

Table 1. Distribution of land holding (% of total) in the Indian Himalayas

Land holding size (ha)	Northeastern Himalayas		Western Himalayas		Total Himalayas		All India	
	Holding	Area	holding	Area	holding	Area	holding	Area
<2	78.4	35.0	87.6	52.6	82.9	41.2	78.3	32.3
2-4	14.6	24.0	9.3	25.9	12.0	25.0	13.3	23.1
4-10	5.5	19.0	2.8	15.7	4.2	17.8	7.3	27.3
>10	1.5	22.2	0.3	5.8	0.9	16.0	1.1	17.5

Table 2. Cultivated and uncultivated crops grown in Uttarakhand

Botanical Name of Plants	English Name	Local Name	Season of Cultivation
Food crops			
Amaranthus caudatus	Chaulai	Ram Dana	K
Cajanus cajan	Pigeon pea	Arhar	K
Echinochloa frumentacea	Barnyard Millet	Madira	K

Eleusine coracana	Finger Millet	Mandua	K
Glycine max	Soyabean	Soyabean/ Bhat	K
Macrotyloma uniflorum	Horse gram	Gahat	K
Oryza sativa	Rice	Dhan	K
Phaseolus vulgaris	Kidney Bean	Rajmah	K
Sesamum indicum	Seasame	Til	K
Setaria italica	Koni	Fox Tail Millet	K
Sorghum bicolor	Jowar	Jowar	K
Vigna mungo	Blackgram	Urad	K
Vigna radiate	Mung	Green gram	K
Vigna sinensis	Cow pea	Lobiya	K
Zea mays	Maize	Makka	K
Brassica alba	White mustard	Rai (white)	R
Brassica napus	Mustard	Sarso	R
Brassica nigra	Black mustard	Rai (black)	R
Cicer arietinum	Gram	Chana	R
Hordeum vulgare	Barley	Jau	R
Lens culinaris	Lentil	Masoor	R
Triticum aestivum	Wheat	Gehu	R
Vegetables and cash crops			
Colocasia escule	Elephant ear fig	Taro	K
Cucumis melo	Muskmelon	Kharbooja	K
Cucumis sativus	Cucumber	Khira	K
Cucurbita maxima	Pumpkin	Kaddu	K
Dioscorea alata	Winged yam	Khamalu	K
Hibiscus esculentus	Okra/Lady fingers	Bhindi	K
Luffa cylindrica	Sponge gourd	Tori	K
Momordica charantia	Bitter gourd	Karela	K
Allium cepa	Onion	Pyaz	R
Allium sativum	Ginger	Adrakh	R
Brassica botrytis	Cauli flower	Gobhi	R
Brassica juncea	Leaf mustard	Rai/Sarson	R
Brassica oleracea (capitata)	Cabbage	Phool gobhi	R
Capsicum annum var. annum	Capsicum	Shimalaa Mirch	R
Curcuma domestica	Turmeric	Haldi	R
Daucus carota	Carrot	Gajar	R
Dolichos lablab	Beans	Sem	R
Lagenaria siceraria	Bottle gourd	Lauki	R

Pisum sativum	Pea	Matar	R
Paspalum scrobiculatum	Kodo Millet	Kodon	R
Solanum lycopersicum	Tomato	Tamatar	R
Solanum melongena	Eggplant	Baigan	R
Solanum tuberosum	Potato	Alu R	
Trigonella foenum-graecum	Fenugreek	Methi	R
Zingiber officinale	Ginger	Adrakh	R
Spinacia oleracea	Spinach	Palak	U
Fodder and Agroforestry plants			
Acacia nilotica	Acacia	Babool	Fr
Aesculus indica	Horse Chest nut	Pankor	Ft
Alnus nepalensis	Nepalese Alder	Utis	Fr
Bambusa arundinacea	Bamboo	Bans	Fr
Celtis australis linn	European hackberry	Khari	Fr
Ficus palmate	Wlid fig	Anijir	Fr
Ficus roxburghii	Elephant ear fig tree	Gular	Fr
Gossypium hirsutum	Cotton	Kapaas	Ft
Grewia optiva	Bheemal	Bhimal	Fr
Melia azedarach	Margosa	Bakain	Fr
Morus alba	Mulbury	Shehtoot	Ft
Pinus roxburghii	Pine	Chir	Ft
Populus androscoggin	Poplar	Poplar	Fr
Prunus cerasoides	Wild himalayan cherry	haya/Paddam	Fr
Quercus leucotrichophora	Blackjack Oak	Banj	Fr
Rubus ellipticus	Yellow Himalayan Raspberry	Lalanchu	Fr
Sapindus emarginatus	Soapnut	Ritha	Ft
Thuja orientalis	Oriental arborvitae	Morphankhi	Ft
Toona ciliata	Toon	Tun	Ft
Ficus spp.	+	Kheena	Fr
Fruit plants			
Carica papaya	Papaya	Papeeta	F
Citrus aurantifolia	Lime	Kagzi nibu	F
Citrus limon	Lemon	Jambiri nimbu	F
Citrus limon	Lime	Nebu	F
Citrus reticulata	Orange	Santara/Narangi	F
Citrus sinensis	Sweet orange	Santara	F
Juglans regia	Walnut	Akhrot	F

Litchi chinensis	Litchi	Litchi	F
Malus sylvestris	Apple	Seb	F
Mangifera indica	Mango	Am	F
Musa paradisiaca	Banana	Kela	F
Prunus armeniaca	Apricot	Zardalu	F
Prunus domestica	Plum	Alu bukhara	F
Prunus persica	Peach	Aadoo	F
Prunus tomentosa	Cherry	Cherry	F
Psidium guajava	Guava	Amrud	F
Punica granatum	Pomegranate	Anar	F
Pyrus spp	Pear	Gola Naspati	F
Pyrus communis	China pear	Nashpati	F
Vitis vinifera	Grapes	Angoor	F

Table 3. *In situ* agro biodiversity conservation on-farm in Uttarakhand

Botanical Name of Plants	English Name	Local Name	Season of Cultivation
Food crops			
Amaranthus caudatus	Chaulai	Ram dana	K
Cajanus cajan	Pigeon pea	Arhar	K
Echinochloa frumentacea	Barnyad Millet	Madira	K
Eleusine coracana	Finger Millet	Mandua	K
Glycine max	Black soyabean	Bhat	K
Macrotyloma uniflorum	Horse gram	Gahat	K
Oryza sativa	Rice	Dhan	K
Phaseolus vulgaris	Kidney Bean	Rajmah	K
Sesamum indicum	Seasame	Til	K
Setaria italica	Fox Tail Millet	Koni	K
Vigna radiata	Mung	Green gram	K
Vigna sinensis	Cow pea	Lobiya	K
Zea mays	Maize	Makka	K
Brassica napus	Mustard	Sarso	R
Brassica nigra	Black mustard	Rai (black)	R
Hordeum vulgare	Barley	Jau	R
Lens culinaris	Lentil	Masoor	R
Triticum aestivum	Wheat	Gehu	R
Vegetables and cash crops			
Celocasia esculenta	Arum	Arbi/Gadari	K

Coriandrum Sativum	Coriander	Dhania	K
Cucumis melo	Muskmelon	Kharbooja	K
Cucumis sativus	Cucumber	Khira	K
Cucurbita maxima	Pumpkin	Kaddu	K
Hibiscus esculentus	Lady`s finger	Bhindi	K
Luffa cylindrica	Sponge gourd	Tori	K
Momordica charantia	Bitter Gourd	Karela	K
Solanum melongena	Brinjal	Bagan	K
Vicia faba	Broad bean	Bakla	K
Allium cepa	Onion	Piyaz	R
Allium sativum	Garlic	Lehsun	R
Brassica botrytis	Cauliflower	Gobhi (Phool)	R
Brassica oleracea (capitata)	Cabbage	Gobhi (patta)	R
Capsicum annum var aanum	Capsicum	Shimla-Mirch	R
Capsicum annum	Chilli	Mirch	R
Curcuma domestica	Turmeric	Haldi	R
Daucus carota	Carrot	Gazar	R
Dolichos lablab	Beans	Sem	R
Lagenaria siceraria	Bottle gourd	Lauki	R
Pisum sativum	Pea	Matar	R
Rasphanus sativus	Radish	Muli	R
Solanum lycopersicum	Tomato	Tamatar	R
Solanum tuberosum	Potato	Alu	R
Spinacia oleracea	Spinach	Palak	R
Zingiber officinale	Ginger	Adrakh	R
Fodder and Agroforestry Plants			
Celtis australis linn	European hackberry	Khari	F
Ficus palmata	Beru	Bedu	F
Ficus roxburghii	Eve's apron	Timal	F
Ficus subincisa	+Chanchri	Chanchri	F
Fraxinus micrantha	Fraxinus	Angu/Litsia	F
Fraxinus xanthoxyloides	Afgahan Ash	Thelka	F
Gossypium arboreum	Thailand cotton	Kapaas	F
Grewia optiva	Not found	Bhimal	F
Lyonia ovalifolia	Not found	Ayar	F
Morus alba	Mulbury	Shehtoot	F
Quercus leucotricophora	Oak	Banj	F
Quescus floribunda	Moru Oak	Moru	F

Rhododendron arboreum	Rose tree	Burash	F
Salix spp.	willow	Karve	F
Thuja orientalis	Oriental arborvitae	Morphankhi	F
Toona ciliata	Red cedar	Tun	F
Fruit Plants			
Citrus aurantifolia	Lime	Kagzi nibu	F
Citrus limon	Lemon	Bara nibu	F
Citrus reticulata	Orange	Santara/Narangi	F
Citrus sinensis	Sweet orange	Malta	F
Juglans regia	Walnut	Akhrot	F
Litchi chinensis	Litchi	Litchi	F
Malus sylvestris	CrabApple	Seb	F
Mangifera indica	Mango	Am	F
Musa paradisiaca	Banana	Kela	F
Prunus armeniaca	Apricot	Khubani	F
Prunus domestica	Plum	Alucha	F
Prunus persica	Peach	Aadoo	F
Pyrus communis	Pear	Gola Naspati	F
Vitis vinifera	Grapes	Angoor	F

R - Rabi season of cultivation- (October to March)
K - Kahrif season of cultivation- (April to Sertember)
F - Fruit species
Fr - Fodder species
Ft - Forest tree
U - Season of cultivation could not be confirmed
+ - English name could not be confirmed

Among the lack of understanding about the farmer's definition of diversity and conservation are also undocumented and hidden costs of the disappearance of knowledge about food systems based on the everyday use of local environments, especially by women. To keep these systems alive is not just the responsibility of the hill women and their families, nor just of the researchers who have long provided these communities their dedicated support, but it is also the responsibility of the civil societies for the sustainable management and development of the Himalaya. Generating understanding and evidence of the connections between diversity in agriculture, the farming practices that support this diversity which is often called as conservation in the scientific parlance is the research theme which needs to be understood by the general ordinance before formulating biodiversity and conservation plans and strategies.

Diversity worth to farmers

In the agricultural systems of Uttarakhand, a diversity of crops and varieties is needed to combat the risks farmers face from small terraced fields, pests, poor moisture, diseases and variations in climate. Consumption of a variety of food crop diversity in this region also indicates the rich knowledge which the community had developed about the properties of these food crops and its impact on human physiology under different climatic conditions, apart from the economic compulsions and developed taste for some of these plants and crops. The knowledgeable farmers consider every plant of the forests important, whether they know its uses or not, but know for sure that it supports many of their useful plants directly or indirectly. Some of the plants are considered locators for identifying a particular location in the forests, while some are considered association partners for healthy growth of certain plants, some are considered as bank reserve hoping to identify some of its uses in future, and the rests are considered useful for soil, water, birds, insects and wild animals. In the context of any agricultural community, documenting all sources of food and fodder is necessary, not only to assess the quantity of food produced per unit of land but also the diversity of food and its availability. A large number of studies show that uncultivated plants and animals are crucial to survival during times of extreme hardship. Knowledge of the uses of available plants and animals can make the difference between those that live and those that die. The value of these cultivated and uncultivated foods in meeting the more regular nutritional needs of vulnerable groups such as young children, lactating mothers and old age people requires proper documentation, understanding and analysis. Our challenge is still to develop an understanding of how to maintain and develop the conditions for ensuring ready access to uncultivated food and fodder sources.

To understand the value of plants and crops variety diversity to the farmers, it would be worth to understand how they manage these resources for example the village women cannot imagine the preparation of food for their family members or the raising of livestock without the leafy greens, grasses, roots and tubers, leaves of trees, bark of trees, fruits, and other items collected from the borders between fields, along roadsides, in small patches of forest or fallow lands, near water springs and other water bodies as well as in the cultivated fields and homesteads. While the diversity of plants in the forests is a consequence of the farmer's ability to know its properties, uses and availability of the desired parts in particular seasons, occurrences in dietary and seasonality, and distance of availability from their village ecosystems. The diversity of crops and crop varieties is a consequence of human choices in close interaction with natural selection. Farmers choose to grow particular crop varieties for the specific qualities they seek, which include production traits such as crop yield and tolerance to pests and diseases, as well as consumption-related attributes such as taste and processing qualities. Since farmers tend to grow varieties based on the traits and attributes they observe rather than those they cannot see, the more sophisticated the construction of the diversity index, often the more obscure is its relationship to the decisions of farmers. Farmers have a variety of techniques for farmer-managed units of diversity, variety of names they have assigned to different landraces, and have their local names based on the morphological characteristics and genetic information.

A number of researches on different components of biodiversity conservation, medicinal plant diversity and other related topics in the Himalayan region (Dhar, 1995; Samant *et al.*, 1998 Maikhuri *et al.*, 1996, 1997) so far have not gone into in-depth research on understanding the farmers concepts and perceptions of diversity and conservation. These researches were largely interrelated in methods and approach, and were based on the scientists understanding of the subject using a particular scientific methodology. There scientific outputs so far have largely conditioned and influenced in building the general understanding of the concepts of diversity and conservation in the Himalayan context. Using this approach, a number of case studies were analyzed across a range of forests, plant species, crops and agricultural economies where crop biodiversity of local economic value remains in the fields of farmers. In this study, researchers have shed more light on questions regarding structures, functions and endemism, and have tried to gather information on "who" maintains diversity, "where" it is maintained, but have not gone into details of "how much" farmers value this diversity as societies and economies change. The findings indicate how factors such as human capital, off-farm income and migration, assets and farm physical conditions influence the way farmers value the attributes of plants and crops varieties. In addition, this study employs higher scales of analysis than previous research on this subject, incorporating the role of social and cultural institutions at the levels of villages, settlements and regions.

In 2004, *Ogal and Phapar* have been reintroduced in the daily rituals of offerings (Prasad) at the Gangotri Temple, a temple dedicated to the goddess Ganga in Uttrakahsi District, situated at 3,200 m asl, nine kilometres downstream of Gaumukh glacier, where the river Ganges originates). Some voluntary organizations are doing a good job to conserve the traditional crop diversity and agricultural genetic variability of this region. For example, Sri Vijay Jardhari and his volunteers from the Jardhar village of Henval valley in Tehri Garhwal district are involved in the collection of seeds of various traditional crops from different parts of the region under their 'Save the Seed' campaign. These experiences though scattered and scanty are ways that demonstrate farmers' economic sustainability, conservation of traditional crop diversity and comparative advantage of the state. The economic advantage of different traditional crops in this mountainous region, is specifically in their being organic products and because they contain better nutrient content as well as tasting better.

Traditional conservation imperatives

The first and foremost principle hypothesis of conservation according to the small farmers is to ensure the availability of basic needs of the communities living in and around natural resources in the rural Himalayan scenarios. Local communities use natural resources more wisely if the link between conservation of resources and improved livelihoods is established. Till date all major developmental and research projects relating to biodiversity and conservation have treated the local people as "passive beneficiaries" and had very little say in the decision making process and had no ownership of the initiatives. Few initiatives had also depended on "de-linking" people from resource base, therefore reduced local stake in the conservation of resources. The links of the social development activities with conservation were often not established and the interventions were input intensive and unsustainable without continuous external support. The uncultivated plants and animals constitute a prominent place in the local food system and are consequently a vital part of agricultural practice.

The economic value of increasing crop productivity through the diffusion of improved, modern varieties has been extensively documented, particularly in the context of industrialized agriculture (e.g., Alston *et al.*, 2000; Evenson and Gollin, 2003). Costs and benefits have also been estimated for plant genetic resources conserved in gene banks, destined principally for use by commercial farmers. In contrast, very less number of researches has sought to understand the value of increasingly scarce, plants from the wild used by the farmers, local varieties of food crops, condiments, spices, and medicinal plants grown by them. This is partly because such varieties are typically found in marginal, isolated environments, where they are traded outside of formal markets. In addition, economists have only recently challenged the commonly-held assumption that local varieties will inevitably be replaced by modern varieties over time (Brush *et al.* 1992; Meng, 1997). Local varieties, often called landraces, generally exhibit high degrees of local adaptation, with particular properties or characteristics that are valuable to the communities in which they are grown and potentially valuable for crop improvement elsewhere, where they may be scarce.

A wide range of cereals, vegetables, pulses, oil seeds, condiments and medicinal and aromatic plants are grown in a village scattered on innumerable piece of agricultural land. A typical farm grows wheat, rice, pulses and vegetable crops each year. The number of uncultivated species used as food is also very large. Some ten to fifteen species of leafy greens associated with agricultural fields, homesteads and common areas were identified by local people as food, mainly herbs, creepers, aquatic plants, shrubs and trees. Uncultivated plants in and around agricultural fields and homesteads are considered "partner plants" in the sense that they grow in association with the cultivated plants. For example, Bathua grows on its own in the wheat or mixed crop mustard field and can be collected when cleaning the agricultural field for the main crop or at any other time. The plants found in the common areas such as along roadsides, boundaries between fields, water bodies, in forest patches, in fallow pastures and around water sources are usually different species, although some grow in several different habitats. The taste, nutritional and medicinal qualities of the same plant are known to be different when collected in agricultural fields compared to uncultivated lands, a result of the dynamic interaction of the soil conditions and the plant genotype.

Uncultivated plants are typically plucked rather than uprooted so that they will regenerate and provide food again. They are in this sense similar to another category of uncultivated foods recognized by local people, that is, cultivated plants that provide edible leaves and shoots in addition to the primary product of cultivation. This includes species such as mustard which is cultivated for the oil seed and jute which is cultivated for its fibrous stalk. The leaves of both plants are also plucked for food. In fact, plucking the leaves in the early part of the growing season is considered by farmers to be good for the plant and is therefore allowed when done with care. These "derived foods" include as well the leaves of pumpkins, sweet gourd, white gourd, and other vegetables grown in the homestead and plucked for food in small quantities almost everyday before the fruit is formed. Flowers of some cultivated plants that are unlikely to be fertilized are collected as well. Around 87% of the households in the rural hill communities have cows or bullocks, usually one or two animals. Almost all households have small ruminants including goats and sheep in the high altitude areas. For example, mid altitude villages have one cow or bullock, while the high altitude villages have more number of sheep and goats apart from cow and bullocks. This is typical not only of farming households but also many landless farmers with homesteads.

Economic change and loss of crop diversity

According to Nautiyal and Kaechele (2007), the maximum area of a single crop cultivated in Uttarakhand is now covered by rice (*O. sativa)* (53 ha), followed by wheat (*T. aesativum)* (49 ha), potato (*S. tuberosum)* (30 ha) and least by *S. indicum* (2 ha). For some time, the scientific communities and the planners have been concerned about declining diversity of crop genetic resources on farms and other types of plant varieties in natural habitat. Many argue that the very processes that engendered the remarkable advances in agricultural productivity during the 20th century, such as the Green Revolution, also eroded the valuable stocks of genetic resources long maintained by farmers. Economists often view the loss of diversity as an unavoidable, unintended consequence of technical change and specialization—a negative externality of progress and development.

As isolated communities have become progressively more linked into global production systems, understanding the impact of economic change on crop diversity has become increasingly important. Although there is little conclusive evidence, it has been hypothesized that processes of economic change and market integration pose some of the greatest threats to crop genetic diversity. With better access to markets and rising incomes, the attributes associated with diverse crop varieties are more easily replaced with purchased inputs and goods. Processes of economic change alter the ways rural people earn their living, removing some of the incentives and knowledge needed to produce diverse crop varieties, thus contributing to genetic erosion. Concerns about the impact of economic change are not new. Starting during the early green revolution period in the 1970s, economists assumed that the introduction of superior seeds would lead farmers to plant all of their crop area with modern varieties. However, landraces of major crops such as rice, wheat, and maize are still grown in a number of places where they outperform modern varieties or have unique traits that farmers value (Smale 2000). Improved varieties are scarce or nonexistent for many so-called "minor" crops that may be of local economic value but not global commercial value. One of the goals of this research is to provide new evidence regarding the impact of economic change on the values that farmers ascribe to their crop genetic resources.

The two faces of economic change

Findings from these and other case studies in the collection indicate that as long as there are harsh production environments where markets are either away or function imperfectly, rural households will continue to depend on the diversity of the materials they grow to ensure the family food supply. This does not, however, mean that those who maintain crop biodiversity need be "left out" of the process of economic development. The relationship of market development and commercialization to crop biodiversity appears to be complex, particularly when considering factors beyond the issue of market access. Many of the case study findings suggest that factors associated with economic development may not, in the short-term, detract from intra crop and intercrop diversity on farms. In some marginal environments, the introduction of modern varieties broadens the range of materials grown rather than narrowing it. Moreover, higher levels of assets often enhance rather than detract from crop biodiversity. On the other hand, diversification in any form is often associated with labor-intensive production. Rising opportunity costs for farm family members in regions undergoing rapid economic change may lead to less diversity within cropping systems. Permanent migration and off-farm employment may ultimately have detrimental effects on crop diversity. These findings underscore an essential point: that there will often be better ways to relieve poverty than through the introduction or diversification of crop varieties. While crop genetic diversity is important to the poor—particularly in terms of meeting food and nutritional requirements—supporting crop genetic diversity conservation is not a way out of poverty per se, unless it can be linked to an income-earning activity.

Shift from traditional to cash crops

The ineffectiveness of research to promote and conserve agro biodiversity and increase the production system of small farmers also contributed to the majority of small farmers to take to commercial crop cultivation in the region. Traditional crops have been sub planted by cash crops in the valley regions and mid altitudes but still traditionally cultivated crops dominates in the high altitudes mean while the production and productivity of these crops are less and they do not meet the food requirement of increasing population as compared to cash crops. Not only the agriculture pattern is shifting but also off season cultivation of vegetables, medicinal plant and fruits are increasing. The main reason for all this is due to high population growth, low yield of subsistence crops, changes in the climatic condition and consequently the situation of food insecurity exits. The practice of traditional farming done in this rain fed hill region is very complicated and called Barahnaja.

Mandua (finger millet), *ramdana/chua* (amaranthus), *rajma* (common kidney beans), o*gal* (buckwheat), *urad* (black gram), *moong* (green gram), *naurangi* (mix of pulses), *gahath* (horse gram), *Bhat* (soybean), *Lobiya* (French beans), *kheera* (cucumber), *bhang* (cannabis) and other crops, are grown together in a mixture which is finely balanced to maximize productivity, preservation of soil fertility, conserving crop-diversity, and is geared towards meeting diverse household requirements (Zardhari, 2000). Spatial extent and cropping intensity in settled farms have not changed as drastically as crop diversity. Crops like *Echinochloa frumentacea, Glycine max, Setaria italica, Panicum miliaceum, Hordeum himalyens, Fagopyrum tataricum* and *Fagopyrum botrydis* have been abandoned or their area drastically reduced due to increasing emphasis on cash crops like *Solanum tuberosum* and *Phaseolus* spp (Saxsena *et al.,* 2005). Unlike some other areas where policies promoted cash crops expansion in the Himalaya seems to be an indigenous initiative driven by a socio-cultural change from subsistence to market economy, comparative ecological advantages for these crops in uplands, changing food habits and supply of food grains at subsidized price by the government (Semwal *et al*., 2003). Farmers are substantially benefited economically by expanding cash crops but at the cost of increased vulnerability to climatic and market uncertainties. Middlemen in the marketing channel exploit marginal farmers. Abandonment of traditional crops/cultivars means a loss of agro biodiversity that remains 'lesser known' or 'unknown' to wider communities. Differing from this widely encountered trend, there are a few isolated pockets where a change from subsistence to market economy progressed without any significant loss of crop diversity. Expansion of potato, whose by-products do not have any fodder value, implies lesser production of fodder from private farmland. Soil loss from the potato field is 6~8 times higher than that from the traditional crops that has been replaced despite of a much higher manure input in the former (Sen *et al.* 1997, 2002). A reduction in fodder yield from farmland and application of larger quantities of manure in the changed scenario implies more threats to forests arising from intensive litter removal and lopping. Plant breeders have developed high yielding varieties of selected crops viz., paddy, wheat and maize, but their adoption has been limited because they are poorly adapted to environmental variability, they require external inputs not accessible to majority of farmers and they yield lower quantities of fodder. The main dimension of agro biodiversity loss in the Central Himalayan agro-ecosystem can be named as Ecological, Social, Policy dimensions of agro-biodiversity loss.

Conclusion

The traditional crop diversity and their landraces in agricultural land use in Himalayas have great significance for long term sustainability of agriculture along with conservation and management practices. As traditional agriculture systems in the mountains is characterized by its dependence on local resources and traditional knowledge systems. In order to optimize food production in these low-input farming systems, farmers knowledge and practices require an institutional policy support for enhancing the value of the traditional crops including the promotion of elite landraces selected on the basis of urban consumption needs.

Since the genetic diversity of crop genetic resources is not fully transparent to the farmers who manage it, individual decisions on the use and management of crop varieties can result in the loss of potentially valuable resources. In developing strategies to sustain agricultural biodiversity in ways that are beneficial to both society and the farmers who manage genetic resources, the greatest benefits and the lowest costs will arise in economic and physical environments in which both the benefits for farmers and the public value of crop diversity are high. However, certain pockets in the region have undergone radical transformation by utilizing traditional crop diversity to their economic advantage. Many farmers' federations and Women's Self Help Groups (SHGs) are cultivating many traditional crops *viz.* buckwheat, horse gram, foxtail millet, barnyard millet, kidney bean etc. at a commercial scale. However, the area under these crops decreased by 80 to 100 percent in the last two decades mainly due to wheat and rice that are available at cheap rates in the public distribution system in comparison to these crops. Very recently, Japan's multinational company Yoshifoomi Kihata that produces baby food has purchased foxtail millet (*mandua*) in bulk from the region. The Women's Cooperative Farmers' Association (RWCF) of Yamuna is marketing bio-products which are now available at retail outlets in metro cities of India. Some products which were consumed traditionally on various religious occasions which because of declining demand were on the verge of extinction are being patronized by state, religious and cultural organizations.

Acknowledgements

Auhours are thankful to the Director, NISTADS, New Delhi for encouragements and providing facilities The funding support from Canadian International Development Agency (CIDA) through Shastri Indo-Canadian Institute (SICI), New Delhi is acknowledged with thanks..

References

Almekinders C J M and Elings A (2001), Collaboration of farmers and breeders: Participatory crop improvement in perspective. *Euphytica,* 122: 425-438.

Alston J M, C Chan-Kang, M C Marra, P Pardey and T J Wyatt (2000), *A meta-analysis of rates of return to agricultural R&D*, Research Report 113, IFPRI Washington, DC.

Brush S B, Taylor J E and Bellon M R (1992), Technology adoption and biological diversity in Andean potato agriculture. *Journal of Development Economic,* 2, 365–387.

Bruin G and Meerman F (2001), New ways of developing agricultural technologies: The Zanzibar experience with participatory Integrated Pest Management. Wageningen University, Wageningen, 167 pp.

Chambers R and Jiggins J (1987a), Agricultural science for resource-poor farmers Part I: Transfer-of-technology and farming systems research. *Agricultural Administration and Extension,* 27: 35-52.

Dhar U (1997), Himalayan Biodiversity: Action Plan. Gyanodaya Publications, Nainital/

Evenson R E and Gollin D (2003), Assessing the impact of the Green Revolution, 1960-2000. *Science*, vol. 300: 758-762. Online document. www.sciencemag.org.

Farooquee Nehal A and R K Maikhuri (2009), Communities and their Agrobiodiversity Priorities for agriculture in Uttarakhand Himalaya, India. *Outlook on Agriculture,* Vol 38, No 4, pp 383–389 383.

Farnworth C, Rozel C and Jiggins J (2003), *Participatory plant breeding and gender analysis. Cali, Colombia*: CGIAR System wide Program on Participatory Research and Gender Analysis. PBB Monograph No. 4. Online document.

Maikhuri R K, Rao K S and Saxena K G (1996), 'Traditional crop diversity for sustainable development of Central Himalayan agroecosystems', *International Journal of Sustainable Development and World Ecology*, Vol 2, pp 1–24.

Maikhuri R K, Rao K S, and Semwal R L (2001a), 'Changing scenario of Himalayan agroecosystems: loss of agrobiodiversity, an indicator of environmental change in Central Himalaya, India', *The Environmentalist*, Vol 21, pp 23–39.

Maikhuri R K, Semwal R L, Rao K S and Saxena K G (1997), 'Eroding traditional crop diversity imperils the sustainability of agriculture systems in Central Himalaya', *Current Science*, Vol 73, No 9, pp 777–782.

Maxted N, B V Ford-Lloyd and J G Hawkes (1997), *Complementary conservation strategies: In: Plant Genetic Conservation: The in situ approach*. Eds. N. Maxted, B.V. Ford-Lloyd and J.G. Hawkes, pp. 15-39. Chapman and Hall.

Meng E C H (1997), Land allocation decisions and insitu conservation of crop genetic resources: The Case of wheat landraces in Turkey. Ph.D. dissertation, University of California at Davis, California.

Nautiyal S, R K Maikhuri, K S Rao, R L Semwal and K G Saxena (2002), Agroecosystem function around a Himalayan Biosphere Reserve. *Journal of Environmental Systems,* 29: 71-100.

Nautiyal Sunil and Harald Kaechele (2007), "Conservation of crop diversity for sustainable landscape development in the mountains of the Indian Himalayan region", *Management of Environmental Quality: An International Journal*, Vol. 18 (5): 514 - 530

Rais Mohammad and Gary W. VanLoon (2006), Executive Summary of NISTADS-SHARP PROJECT and Proceedings of Dissemination Workshop on Agrobiodiversity Conservation for Sustainable Livelihood of Hill Communities in Uttaranchal. NISTAD and Queen's University.

Roling N G (1996), Towards an interactive science. *European Journal of Agricultural Education and Extension*, 2:35-48.

Samant S S, U Dhar and R S Rawal (1998), Biodiversity status of a protected area of west Himalaya – Askot Wildlife Sanctuary. *International Journal of Sustainable Development and World Ecology*, 5:193-203.

Saxena K G, Maikhuri R K and Rao K S (2005), Changes in agricultural biodiversity: Implications for sustainable livelihood in the Himalaya. *Journal of Mountain Science*, 2, 23-31.

Semwal Maikuri and Rao (2001), *Agriculture, Ecology, Practices and Productivity, Garhwal Himalaya: Nature, Culture and Society*, Kandari O P and Gusain O P (eds), Transmedia, Srinagar Garhwal , pp 259-275.

Sen K K, K S Rao and K G Saxena (1997), Soil erosion due to settled upland farming in Himalaya: A case study in Pranmati. *International Journal for Sustainable Development and World Ecology*, 4:65-74.

Smale M, ed. (2006), Valuing Crop Biodiversity: On Farm Genetic Resources and Economic Change, Wallingford, UK:CABI Publications

Zardhari V (2000), Barahanajȧ – Twelve Food Grains: Traditional Mixed Farming System. *Leisa India*, Vol. 2 (3): 25-8.

Olea ferruginea Royle: A potential tree crop for sustainable development of north west Himalaya

S.C. Joshi

G.B. Pant Institute of Himalayan Environment and Development, Garhwal Unit, P. Box No. 92, Srinagar-Garhwal, 246 174, Uttarakhand, India
[E-Mail: joshisc@hotmail.com]

Introduction

The Indian Himalayan region is a well known repository of several ecologically and economically important plant species. However, continually increasing human and livestock population, and unscientific way of their exploitation are posing a great threat to these natural resources. As a result, some of the species have found place in the Red Data Book under various categories of threatened plants (Samant, Dhar and Palni, 1998). This is being further exacerbated by global climate change. It is widely accepted that conservation and judicious use of indigenous/native plant species is vital for the ecological and environmental stability of any region as the use of exotic plant species has been reported to result in considerable negative consequences in the long-term (Kamaljit *et al.*, 2007). The olive, known as jaitoon in Hindi or Sanskrit, is one of the oldest and most important agricultural fruit tree crops. It is native to the Mediterranean region, tropical and central Asia and various parts of Africa. It belongs to the family oleaceae, which includes about 30 genera including *Olea* and 600 species. It is regarded as holy plant to symbolize peace and abundance. Ancient Greeks and Romans consecrated it to the Goddess of Wisdom. The victors of Olympic Games in ancient Greece were awarded with crown made of olive branches (Gordini, 1993, Rhizopoulou, 2004). The genus *Olea* includes more than 30 species world over (Anonymous, 1997). These species have been further subdivided into four subspecies or stocks namely *O. europaea* L. ssp. *sativa* Hoffm. et Link, the most cultivated olive tree in the world, *O. europaea* L. ssp *sylvestris* Mill (syn. *O. oleaster* Hoffm. et Link), a wild type, *O. europaea* L. ssp *laperrini* Batt. et Trab., a native to Africa, and *O. europaea* L. ssp *ferruginea* Royle (Syn. *O. cuspidata* Wall. ex. G. Don), a native to the Himalaya (Brousse, 1987). However, there are conflicting reports on the phylogenetic relationships between cultivated olives and related wild forms (Bitonti, 1999). There are thousands of cultivars of the olive, but only a few are grown to a large extent. It is a long lived plant and has been cultivated for oil extraction and olive fruit for at least 6000 years (Zohary, 1994, Vossen, 2007). Both olive products such as table olives and olive oil have become important commodities in the world market because of their nutritional values, and the demand for these products is increasing rapidly. As a result, the cultivation of olive is expanding to other parts of the world (Sedgley, 2000, Rugini, 1986). *O. europaea* was also introduced and experimentally grown in India in the fifties. A number of olive varieties/cultivars were introduced in to India under the "Indo- Italian Programme for the Development of Temperature Climate Fruit Crops" (Prevost *et al.*, 1993). However, most of these varieties/cultivars are now not producing fruits in Himachal Pradesh.

About six species of *Olea* occur in India, one of them is *O. ferruginea* (Anonymous, 1997). Besides in India, it is also found in Afghanistan, Nepal and Pakistan. *O. ferruginea*, Indian olive, is a multipurpose, zero waste, evergreen tree species as its different parts are used for different purposes. It yields a number of useful products including fodder, fuelwood and timber. The fodder is very nutritional and increases the production of milk in the cattle. The leaves and bark of *O. ferruginea* are used as antiperiodic in fever and debility (Anonymous, 1997). The leaves are also used in toothache, astringent medicines, mouth ulcer, demulcents, gonorrhoea, sore throat, hoarse voice, and as a tea beverage, mild digestive aid, antiseptic and anthelmintic (Ahmad *et al.*, 2002, Zabihullah *et al.*, 2006). The fruits are edible, pickled, and used as appetizers, emmenagogues, antidiabetics and a source of olive oil, and they are also used in typhoid, jaundice, biliousness, and scabies, burning of the eyes, toothache and caries of the teeth (Zabihullah *et al.*, 2006, Ahmad, 2007). The oil is useful in cooking, rheumatism, joint pains, malaria, gonorrhoea, skin diseases and cosmetics (Ahmad, 2007, Ahmad *et al.*, 2006). The root is used in asthma, scorpion sting, rheumatism and headache (Hussain *et al.*, 2006). The timber is used for making agricultural implements. The fruits and leaves of olive contain a very high amount of polyphenols and antioxidants which are credited with many health benefits (Briante, 2002, Martinez-Gonzalez *et al.*, 2002, Braun, 2005, Gill *et al.*, 2005). *O. ferruginea* has also been reported to contain oleanolic acid, a biologically very important compound, involved in various activities such as antitumour, antioxidant, anti-inflammatory, antipruritic, antiallergic, antiviral, and immunomodulatory (Sultana and Ata, 2008).

In India, *O. ferruginea* is found in western Himalaya from Kashmir to Kumaun up to an altitude of 2400 m (Anonymous). It is locally known as Kahu (Kao) in Himachal Pradesh and Bairbanj in Garhwal. This species is mainly being used as fodder and fuelwood, and as a rootstock in grafting experiments with *O. europaea* in Himachal Pradesh. The present paper discusses some important features of *O. ferruginea,* which point towards that *O. ferruginea* is a potential tree for rehabilitation of degraded lands in the north west Himalaya and a source of olive oil.

Material and methods

Study site and plant material

O. ferruginea is a medium sized tree up to 15 m in height and occurs chiefly along the outer hills and in the inner dry valleys of western Himalaya. It is occasionally cultivated in the plains of north India. Young shoots, petioles and leaves beneath are covered with reddish scale. Leaves are about 4-10 cm in length, opposite, oblong-lanceolate, entire, acute or accuminate, coriaceous, dark green and shining upper surface, lower densely covered with minute red brown scales. It flowers in the months of May and the flowers are 3-7 mm long, white, bisexual, in numerous short axillary panicles. Its fruiting period is August-September. Drupe is 7-8 mm long, ovoid, black but green when unripe (Anonymous, 1997, Dhaliwal and Sharma, 1999).

Seasonal changes in sclerophylly and leaf water status

This study was conducted on plants growing in the rainfed agricultural fields at Kolibehar area in Kullu district (altitude 1200 m amsl, latitude 31°38'N and longitude 77°60'E) of Himachal Pradesh. The site has a subtropical to temperate alpine type climate. 5-6 trees of *O. ferruginea* of almost equal age and height growing in the rainfed fields were randomly selected for the present investigation. No lopping was allowed in the selected trees during the experiment. Youngest fully mature leaves from the apical regions of 5-6 different plants were harvested between 1100 and 1300 h and kept in the airtight plastic bag placed in an icebox and brought immediately to the laboratory for measurements of leaf mass per unit area (the ratio of leaf dry mass to leaf area), density of foliar tissue (the ratio of leaf dry mass to fresh mass), succulence (the ratio of difference in leaf fresh and dry mass to leaf area), water content at saturation (the ratio of difference in leaf saturated mass and fresh mass to dry mass), leaf water potential, relative water content (RWC), and chlorophyll concentration in terms of SPAD values. These measurements were made using the same plants between the 15th and 20th day of every month from May to October (active growth period of the plant) to determine their seasonal patterns. For instantaneous assessment of chlorophyll concentration, SPAD values using Minolta SPAD-502 leaf chlorophyll meter were recorded at 2-3 different places for each leaf and the values were averaged. Water potential was measured using WP4 Dew Point Potentiameter (Decagon Devices, Inc., USA). The indices of sclerophylly (leaf mass per unit area and density of foliar tissue), and of leaf water status (RWC, succulence and water content at saturation) were calculated according to (Groom and Lamont, 1999) and (Bacelar *et al.*, 2004), respectively. Leaf area was measured using a high precision digital Planimeter (Placom, Japan). The environmental parameters like average monthly temperature and rainfall data were recorded during the experiment. The data were analyzed statistically following Mather (Mather, 1973).

Extraction of oil and fatty acid composition determination

For extraction of olive oil and determination of fatty acid composition, four different populations of *O. ferruginea*, three in Kullu district (Saioond, Kolibehar and Kais) and one in Mandi district (Thalaut) were selected as the source of fruits and seeds. These sites differed in terms of geographical characteristics such as altitude, latitude, longitude and aspects Table 1. Phenotypically superior plants of *O. ferruginea* were selected on the basis of morphological characteristics such as health and growth. Fully ripened fruits were collected from the selected trees of the four populations and soaked in water for 24 hr in order to soften the pulp. The fruits were then washed thoroughly under running tap water to remove the pulp (the exocarp and the mesocarp of the fruit) from the fruits. The fruits and seeds were air dried for 7 days at room temperature, and then used for the determination of olive oil content and fatty acid composition following the standard techniques (Joshi 2012).

Table 1. Geographical characteristics of different selected populations

Population	Altitude (m)	Aspect	Latitude	Longitude
Thalaut	957	North East	31°42′53″ N	77°12′ 06″ E
Kolibehar	1160	South East	31°54′ 95″ N	77°07′ 57″ E
Saioond	1220	East	31°54′ 41″ N	77°09′ 51″ E
Kais	1422	West	32°01′ 23″ N	77°08′ 18″ E

Results and discussion

Water stress is one of the major environmental factors affecting survival, growth, productivity and geographical distributions of plants worldwide. Plants growing in the rainfed areas of north west Himalaya experience repeated water stress periods, depending on rainfalls, which limit growth and productivity. Over the years, this problem has been further aggravated as the rainfall pattern in the region has become highly variable and unpredictable due to global climate change. This has resulted in severe land degradation problem in the north west Himalaya (Maikhuri and Rao, 2002). It has been documented that for better growth and productivity under rainfed conditions plants should possess strategies that enable them to maintain higher photosynthetic carbon assimilation. Identification of such strategies that make plants to grow in water stress conditions is thus helpful in the selection of plant species suitable for rainfed areas. The olive tree (*Olea europaea* L.) is amongst the most abundant cultivated species in the Mediterranean region. It is known to possess adaptive mechanisms to tolerate severe water stress (Larcher *et al.*, 1981). However, water stress has been shown to impact fruit development and composition in olive trees (Gucci *et al.*, 2009). Further, the timing and the degree of water stress have been reported to influence yield, fruit size and properties of oil (Lavee *et al.*, 2007). Thus, failure of fruit production in the introduced varieties / cultivars of O. *europaea in* Himachal Pradesh could be partly attributed to water stress. On the other hand, under the similar climatic conditions *O. ferruginea* shows fruit production, indicating its better water stress tolerating capacity.

The average monthly temperature and total monthly rainfall data of the study site during the experimental period are given in Fig. 1. The average monthly temperature during the experimental period varied from 21 to 26°C. The area was characterized by monthly variation in rainfall, with maximum rain occurring during rainy season (July- September) and minimum during summer (May and June) and autumn season (October). Thus the plants growing there are exposed to repeated water deficit periods. Studies on morphological, physiological, biochemical and molecular adaptations of plants to water stress have shown that different species may employ different survival strategies under field conditions (Auge, 1998). Several studies have shown that some species respond to adverse environmental conditions with an increase in sclerophylly (Bacelar, 2004). It is evident from the data Table 2 that *O. ferruginea* possesses very high leaf mass per unit area and density of foliar tissue. These values were significantly higher when compared with the data of other co-occurring species such as *Grewia optiva*, *Robinia pseudoacacia*, *Ulmus wallichiana* and *W. villosa* (data of other co-occurring species are not given). The leaf mass per unit area of *O. ferruginea* varied from a minimum of 199.34 g m^{-2} in July to a maximum of 210.66 g m^{-2} in August and density of foliar tissue from a minimum of 0.47 in September to a maximum of 0.53 mg g^{-1} in July. This shows that *O. ferruginea* exhibits a higher degree of sclerophylly than the other co-occurring species. These leaf attributes indicate its greater ability to survive drought conditions (Mediavilla S, 2001). These leaf attributes did not show significant seasonal variations, probably indicating inherent nature of these leaf attributes. Further, succulence index of *O. ferruginea* was significantly greater than that of other species. Succulency is known to help plants to survive in arid conditions as they are rich in water content per surface unit (Bacelar, 2004). On the other hand, O. *ferruginea* showed very low water content at saturation that ranged from 0.24±0.04 to 0.28±0.06 g H_2O g^{-1}dry mass. This value was lower compared to that of other co-occurring species, indicating its ability to withstand arid environments (Bacelar *et al.*, 2004).

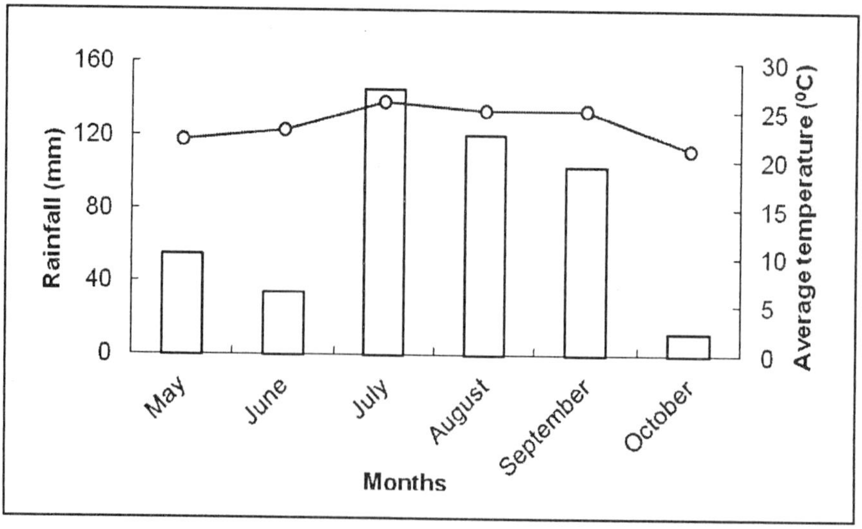

Fig. 1. Total monthly rainfall (bar) and monthly average temperature (line) data of the experimental site.

Table 2. Seasonal courses of leaf mass per unit area, density of foliar tissue, succulence and water content at saturation in leaves of *O. ferruginea* during May to October

Parameters	May	June	July	August	September	October
Leaf mass per unit area (g m^{-2})	208.78±19.63	199.36±16.35	199.34±14.75	210.66±15.38	203.78±16.51	206.89±14.07
Density of foliar tissue (g g^{-1})	0.50±0.05	0.50±0.03	0.53±0.01	0.51±0.02	0.47±0.03	0.51±0.02
Succulence (mg H_2O cm^{-2})	20.75±1.66	19.44±1.36	19.88±1.29	19.62±1.33	18.99±1.50	18.87±1.33

O. ferruginea showed the lowest leaf water potential and RWC as compared to the other species (data not shown). It has been demonstrated that the minimum seasonal water potential of a plant reflects the maximum water deficit that it must tolerate to maintain physiological activity (Bhaskar and Ackerly, 2006). The maintenance of low leaf potential and RWC by *O. ferruginea* compared to other species during its active growth period indicates its greater tolerance to drought (Auge *et al.*, 1998). In many plant species, photosynthesis is greatly sensitive to leaf water potential (Hsiao, 1973). Though photosynthesis could not be measured in the present investigation, chlorophyll concentration measured in terms of SPAD values was significantly higher in *O. ferruginea* during the active growth period. This might enable *O. ferruginea* plant to have and maintain high photosynthetic activity during the growing season and hence higher growth and productivity compared to other species as a strong correlation between leaf nitrogen and chlorophyll contents and in turn with photosynthetic rate and hence growth has been reported in several plants species (Filella *et al.*, 1995).

The fruits and seeds of *O. ferruginea* contain olive oil. The populations of *O. ferruginea* differing in respect to altitude, latitude and longitude also differ in their oil content. The olive oil content was considerably higher in the fruits than in the seeds. The olive oil contents in the fruits and seeds, depending on populations, varied from approximately 21 to 27% and 8 to 13%, respectively Fig 1. The olive oil contents in the world's most prominent olive oil varieties have been reported to vary from a minimum of 13% to a maximum of 28%[9]. This shows that the olive oil content in the fruits of Thalaut population is on par with that of the world's most prominent olive oil cultivar. Further, in all the populations the pulp contains the highest content of olive oil as the seed in the present study also includes endocarp.

Several studies have shown that the contents of olive oil and fatty acid composition may be greatly influenced by several factors such as cultivar/variety, pedoclimatic conditions, agronomic treatments, harvesting time, season, storage and the extraction processes used (Sedgley, 2000, Moussaoui *et al.*, 2008). The differences in the olive oil contents of fruits and seeds among the four populations in the present study may be partly attributed to the differences in edaphic and climatic conditions of the sites as the sites differ in respect to the aspects, altitude, longitude and latitude.

Besides in oil contents, differences were observed in the fatty acid composition between the fruit and seed oils Figs. 2-5. It is well known that the profile of fatty acids varies greatly in vegetable oils and they play an important role in human diets and health. Several Dietary Advisory Bodies recommend balanced intake of saturated fatty acids, monounsaturated fatty acids and polyunsaturated fatty acids (Leon *et al.*, 2004). The seed oil in general contained higher contents of polyunsaturated fatty acids (PUFA) and unsaturated fatty acids (UFA) as compared to the fruit oil Figs. 3b and 4b. The levels of PUFA and UFA in the seed oils varied from 18.1 to 22.3% and 85.8 to 88.2%, respectively. Compared to the fruit oils, the seed oils exhibited considerably lower contents of saturated fatty acids Fig. 4a. The ratio of UFA to saturated fatty acids (SFA) was also higher for the seed oil; the ratio varied from 6.52 to 7.48 in the seed oil Fig. 5a. The contents of oleic acid, monounsaturated fatty acids (MUFA), and MUFA/PUFA ratio were almost equal in the oils from the fruit and seeds Figs. 2b, 3a and 5b. Variations in the fatty acid composition and contents of the oils extracted from different parts of olive fruit have been reported (Moussaoui *et al.*, 2008, Bianchi *et al.*, 1994).

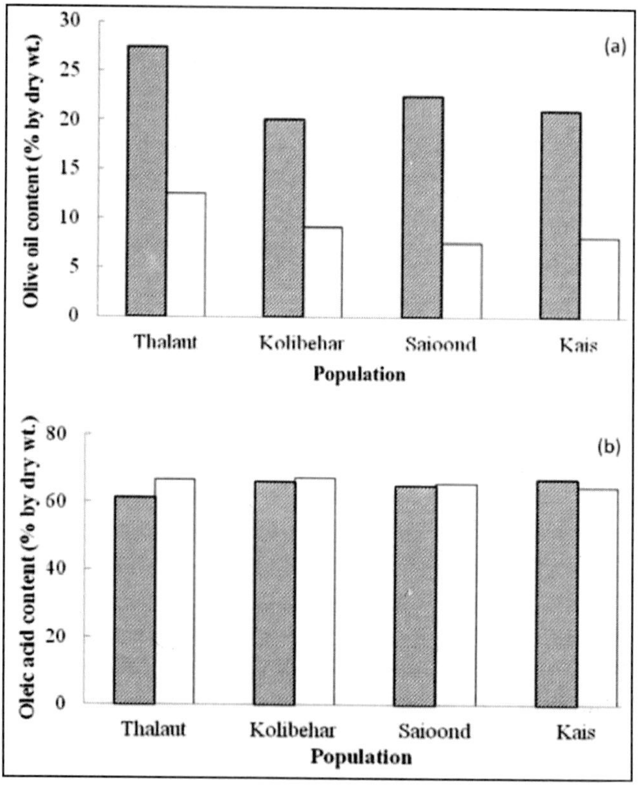

Fig. 2. Olive oil (a) and oleic acid (b) content in the fruit ( ) and seeds ( ) of four different populations of O. *ferruginea.*

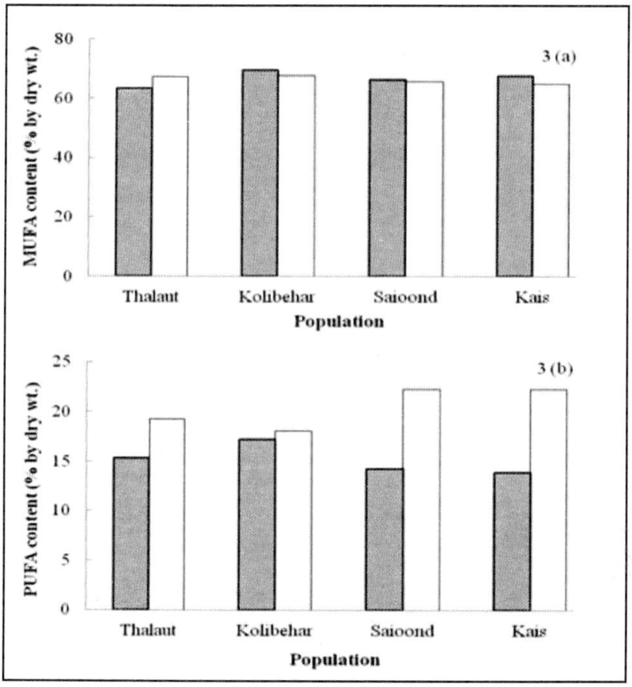

Fig. 3- Monounsaturated fatty acid (MUFA) (a) and polyunsaturated fatty acid (PUFA) (b) content in the fruit ( ) and seeds ( ) of four different populations of O. *ferruginea.*

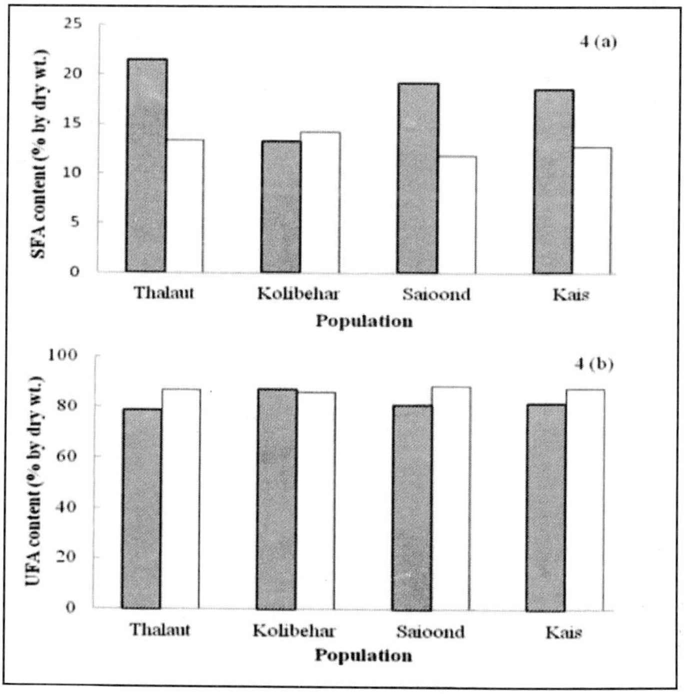

Fig. 4- Saturated fatty acid (SFA) (a) and unsaturated fatty acid (UFA) (b) content in the fruit ( ) and seeds ( ) of four different populations of O. *ferruginea.*

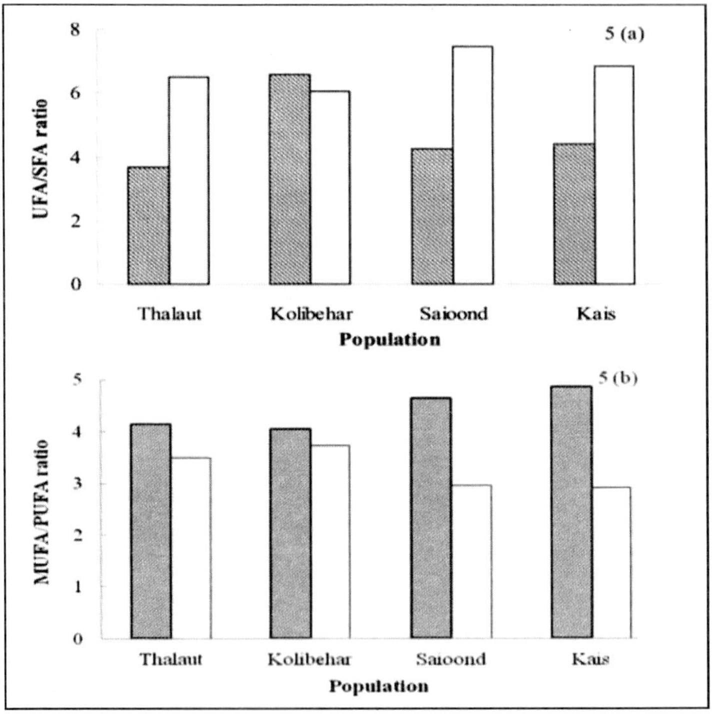

Fig. 5- Ratio of UFA/SFA (a) and MUFA/PUFA (b) in the fruit ( ) and seed ( ) of four different populations of O. *ferruginea.*

To conclude, it could be suggested based on the data that *O. ferruginea* is a potential tree for afforestation of degraded lands as it possesses different mechanisms to survive and grow under drought conditions prevailing in the rainfed areas and thus could help in combating land degradation problem prevailing in the Himalayan region. It may also be a potential source of olive oil. Thus, sustainable use of this potential of the tree crop could be useful in the sustainable development of the areas where it grows.

Acknowledgement

Author is thankful to the Director, G.B. Pant Institute of Himalayan Environment and Development, Kosi-Katarmal, Almora for providing the necessary facilities and encouragement for this work.

References

Anonymous, *The wealth of India. Raw materials. vol. VII*, (Council of Scientific and Industrial Research, New Delhi) 1997, pp. 330.

Ahmad H, Ahmad A and Jan M M, The medicinal plants of Salt Range, *J Biol Sci.*, 29 (2002) 175-177.

Ahmad S S, Medicinal wild plants from Lahore-Islamabad motorway (M-2). *Pak J Bot.*, 39 (2007) 355-375.

Ahmad S, Ali A, Bibi S, Marwat K B and Hassan G, Ethnobotanical study on some medicinal plants of Ouch district lower Dir, Pakistan. *Pak J Pl Sci.*, 12 (2006) 65-71.

Auge R M, Duan X, Crocker J L, Witte W T and Greon C D, Foliar dehydration tolerance of twelve decidous tree species. *J Exptl Bot.*, 49 (1998) 753-759.

Brousse G, Olive. In: *Oil crops of the world: their breeding and utilization,* edited by Robbelen G. *et al.*, (McGraw- Hill Publishing Company, New york, USA) 1987, pp. 462-474.

Bitonti M B *et al.*, Amount and organisation of the heterochromatin in *Olea europaea* and related species. *Heredity,* 83 (1999) 188-195.

Briante R, La Cara F, febbraio, F, Batumi M and Nucci R, Bioactive derivatives from oleuropein by a biotransformation on *Olea europaea* leaf extracts. J Biotechnol., 93(2002) 109-119.

Braun L, Olive-leaf extract. *J Comp Med.,* 4 (2005) 69-73.

Bacelar E A, Correia C M, Moutinho-Pereira J M, Goncalves B C, Lopes J I and Torres-Pereira J M G, Sclerophylly and leaf anatomical traits of five field-grown olive cultivars growing under drought conditions. *Tree Physiol.*, 24 (2004) 233-239.

Bhaskar R and Ackerly D D, Ecological relevance of minimum seasonal water potentials. Physiol Plant., 127 (2006) 353-359.

Bianchi G, Vlahov G, Composition of lipid classes in the morphologically different parts of olive fruit (*Olea europaea* L.). *Fat Sci Technol.*, 96 (1994) 72-77.

Dhaliwal D S and Sharma M, *Flora of Kullu district,* (Bishen Singh Mahendra Pal Singh, Dehradun) 1999, pp. 744.

Filella I, Serrano L, Serra J and Penuelas J, Evaluating wheat nitrogen status with canopy reflectance indices and discriminant analysis. *Crop Sci.*, 35 (1995) 1400-1405.

Gordini A, *Olive Growing,* (Agrotech SpA, Rome) 1993, pp. 54.

Gill C I R, Boyd A, McDermott E, McCann M, Servili M, Selvaggini R, Taticchi A, Esposto S, Montedoro G-F, McGlynn H and Rowland I, Potential anti-cancer effects of virgin olive oil phenols on colorectal carcinogenesis models in vitro. *Int J Cancer* 117 (2005) 1-7.

Groom P K and Lamont B B, Which common indices of sclerophylly best reflect differences in leaf structure? *Ecoscience* 6 (1999) 471-474.

Gucci R, Lodolini, E M and Rapoport H F, Water deficit-induced changes in mesocarp cellular processes and the relationship between mesocarp and endocarp during olive fruit development. *Tree Physiol.,* 29 (2009) 1575-1585.

Hussain F, Badshah L and Dastagir G, Folk medicinal uses of some plants of south Waziristan, Pakistan. *Pak J Pl Sci.*, 12 (2006) 27-39.

Hsiao T C, Plant responses to water stress. *Ann Rev Plant Physiol.*, 24 (1973) 519-570.

Joshi S, *Olea ferruginea* Royle, Indian olive: an underutilised fruit tree crop of north-west Himalaya. *Fruits*, 67 (2012) 121-126.

Kamaljit K, Jalota, R K and Kohli R K, Social acceptance for ecological services of exotic *Eucalyptus tereticornis* and indigenous *Dalbergia sissoo* plantations in N-W India. *Int J Ecol Econ Stat.*, 7(2007) 68-85.

Larcher W, Moroes, J A P V and Bauer H, Adaptive responses of leaf water potential, CO_2 gas exchange and water use efficiency of *Olea europaea* during drying and rewatering, in: *Components of productivity of Mediterranean-climate regions, basic and applied aspects*, edited by Margaris, N S and Mooney, H A (Dr. W. Junk Publishers, The Hague) 1981, pp. 77-84.

Lavee S, Hanoch E, Wodner M and Abramowich H, The effect of predetermined deficit irrigation in performance of cv. Muhasan (*Olea europaea* L.) in the eastern coastal plain of Israel. *Sci Hortic.*, 112(2007) 156-163.

Leon L, Uceda M, Jimenez A, Martin L M and Rallo L, Variability of fatty acid composition in olive (*Olea europaea* L.) progenies. *Spanish J Agric Res.*, 2 (2004) 353-359.

Martinez-Gonzalez M A, Fernandez-Jarne E, Serrano-Martinez M, Marti A, Martinez J A and Martin-Moreno J M, Mediterranean diet and reduction in the risk of a first acute myocardial infection: An operational healthy dietary score. *Eur J Nutr Aug.*, 41(2002) 153-160.

Mather K, *Statistical analysis in biology*, (Chapman and Hall, London) 1973, pp. 267.

Maikhuri R K and Rao K S, Rehabilitation of degraded land: Land use and land cover change impacts and strategies in the Indian Himalayan mountains. *IHDP News letter update* 2 (2002) 18.

Mediavilla S, Escudero A and Heilmeier H, Internal leaf anatomy and photosynthetic resource-use efficiency:interspecific and intraspecific comparisions. *Tree Physiol.*, 21(2001) 251-261.

Moussaoui R, Labbaci W, Hemar N, Youyou A and Amir Y, Physico-chemical characteristics of oils extracted from three compartments of the olive fruit (pulp, endocarp and seed) of variety Chemlal cultivated in Kabylia (Algeria). *J Food Agric Env.*, 6 (2008) 52-55.

Prevost G, Bartolini G and Messeri C, *Cultiver di olive Italiane e loro sinonimi*, (Menegozzo, ed. Lucca) 1993, pp. 67.

Rhizopoulou S, Symbolic plant (s) of the Olympic Games. *J Exptl Bot.*, 55 (2004)1601-1606.

Rugini E, Olive, in: *Biotechnology in agriculture and forestry: Trees I*, edited by Bajaj Y P S, (Springer Verlag, Berlin) 1986, pp. 253-276.

Samant S S, Dhar U and Palni L M S, *Medicinal plants of Indian Himalaya: Diversity distribution potential value*, (Gyanodaya Prakashan, Nainital) 1998, pp.476.

Sultana N and Ata A, Oleanolic acid and related derivatives as medicinally important compounds. *J Enzyme Inhibition Medicinal Chem.*, 23 (2008) 739-756.

Sedgley M, *Wild olive selection for quality oil production*. (Rural Industries Research and Development Corporation, Kingston Act) 2000, pp. 11.

Vossen P, Olive oil: History, production and characteristics of the World's classic oils. *HortSci.*, 42 (2007) 1093-1100.

Zohary D, The wild genetic resources of the cultivated olive. *Acta Hort.*, 356 (1994) 62-650.

Zabihullah Q, Rashid A and Akhtar N, Ethnobotanical survey in Kot Manzaray Baba valley Malakand agency, Pakistan. *Pak J Pl Sci.*, 12 (2006) 115-121.

Climate change impacts in central Himalayan agriculture: Integrating local perception and traditional knowledge for adaptation

R.K. Maihuri & L.S. Rawat*

G. B. Pant Institute of Himalayan Environment and Development, Post Box: 92, Upper Bhaktiyan, Srinagar, Garhwal-246 174, Uttarakhand, India
*[rkmaikhuri@rediffmail.com]

Introduction

The mountain region of the central Himalaya is highly sensitive and vulnerable to climate change. The intergovernmental panel on climate change (IPPC) report have confirmed that we are now locked into inevitable changes in climate. Changes on climate such as mean temperature, sea level rise and increase frequency and intensity of extreme weather events as well as changes in precipitation patterns have been projected. Although there is a lack of certainty in predicting and quantifying climate change impacts on mountain agriculture and rural socio-economic systems, it is well known that climate impacts threaten a major dimension of human well being, namely food and nutritional security. The food security of the people inhabited here depends heavily on climate sensitive sectors (agriculture, livestock, forestry and their inter-linkages and inter-dependencies with each other) and the impact of climate change has the potential to break-down socio-economic development as well as livelihood support systems linked to all these sectors. The forest is a common resources pool and how well it is managed will determine how less vulnerable community livelihoods and development efforts will be to climate change impacts. The rural landscape of the central Himalaya has close proximity to the forest ecosystems which provide livelihood to the local people inhabited there and it likely to be impacted by climate change in various ways as stated in the forth assessment report of Intergovernmental Panel on Climate Change (IPPC, 2007). In addition, the National Action Plan for Climate Change (NAPCC) under National Mission for Sustaining the Himalayan Ecosystem (NMSHE) strongly highlighted this issue.

The population of the mountains continues to grow at high rates, while the extent of agricultural areas has stagnated or is decreasing in some potential grain producing areas due to introduction of developmental power project, expansion of industrial growth in Tarai belt etc. During recent past due to increase in climatic variability, agriculture productivity has deteriorated and it has directly affected the food security of approximately 70% of rural people inhabited in the marginal mountain areas. To feed everyone adequately, agricultural production will have to increased many fold in near future. Extreme droughts in the year of 2008 and 2009 and variability in rainfall between 2000 to 2010-12 have often led not only to abandonment of rainfed (marginal) cultivable land, but also to an exacerbation of poverty through the degradation of the bio-resources leading to large scale out-migration of people in search of job.

However, research in mountain agriculture sector has focused for some decades more on varietals test and release of new varieties with improving productivity of few selected crops i.e. wheat and rice only confined to irrigated land but crops grown in rainfed agriculture is still ignored to the large extent. Now much focused research need to cope with adverse and irregular climatic conditions including rainfall variability or shifting weather patterns particularly in central Himalayan mountainous region. Climate change will have a direct effect on water resources, putting increased stress on water availability for irrigation. Furthermore, diseases and insect populations are strongly dependent upon temperature and humidity and changes could alter their distribution and virulence.

Accepting that global change occurs at rates unprecedented in recorded human history for which people inhabited in central Himalayan region need to develop adaptation mechanism and coping strategies in agriculture sector, so as to ensure sustainable livelihoods for people living in this region (Saxena *et al.*, 2004). However, during recent past local communities have already adopted some best practices as a means of adaptation and mitigation due to alteration in rainfall pattern. Therefore, enhancing adaptive capacity is a necessary condition for reducing vulnerability, especially for the most vulnerable regions and socio-economic groups. These groups are already under pressure due to population growth, resource depletion, and socio-economic inequalities.

The present study intends to document local people perception and knowledge in the context of climate change, general information on traditional agriculture in agrobiodiversity, vulnerability to climate change particularly impact of drought/shift of rainfall, insect/pest incidence on agriculture and suggest appropriate climate change adaptation strategies.

Materials and methods

Description of the traditional mountain agroecosystem in central Himalaya

Central Himalaya has a long heritage of subsistence economy, with agriculture being the core component, in which over 70% of the people are involved. The variations in altitude, topography, climate, availability of forest resources and water for irrigation and socio-economic and cultural factors exist producing a variety of land use patterns in the region. The heterogeneity becomes more complex when ecological conditions are superimposed. The traditional agro-ecosystems of this region exhibits a great deal of variability in crop diversity, crop composition, crop rotation etc. along an altitudinal transect due to corresponding variations in a number of factors which influence agricultural practices. Rain-fed and irrigated land use systems are important in this region in which the former is the predominant form and covers almost 89% of the total agricultural land of the area (Maikhuri *et al.*, 1997, 2001, 2009). Mixed and mono-cropping, particularly in rainfed agro-ecosystem practiced on slopy terraces, has an extremely long tradition in this region. The cropping patterns generally upto 1800 masl, and sometimes 2000 masl are built around two major cropping seasons viz., Kharif (April-October) and Rabi (October-April). Traditional farmers of this region generally cultivate 10 to 12 and some times more crop species in rainfed agro-ecosystem to meet all their food requirements throughout the year. The majority of the crops cultivated by them are traditional or under utilized and include *Amaranthus* spp, *Hordeum himalayens*, *Eluesine coracana, Fagopyrum spp., Setaria italica, Echinochloa frumentacea, Macrotyloma uniflorum, Vigna unguiculata, Parilla frutense*, etc. From rainfed agriculture generally three crops are taken in every two years while from irrigated land two crops are taken in each year. However, sometimes in a few localities at lower altitude, a third crop of hog millet (*Panicum milliaceum*) is also taken between May-June on an small scale from irrigated land. In some specific localities of higher altitude an unique variety of wheat is cultivated which takes as much as 12 months (September – August) to grow and to be harvested. Contrary to the rainfed agriculture of lower and middle altitudes, where the land under mixed cropping of finger millet and pulses in the *Kharif* season remains fallow during Rabi season for 6 months, at higher altitude (>2300 m asl) it remains fallow after the Rabi season crops for a longer (10 months) period.

Traditional crops and their unique characteristics and potential

Varied edaphic, topographic and climatic factors as well as different selection pressures over centuries of cultivation resulted in immense variation (Arora, 1987; Maikhuri *et al.*, 1996, 1998, 2001, 2009). Indigenous varieties evolved over a span of centuries and are adapted to particular areas. The old varieties of traditional crops/millets/pseudomillets, pulses, wheat/rice etc (usually called primitive cultivars or land races) withstood the rigors of time, escaped attacks from insects, pests, diseases and tolerated harsh climatic conditions. They posses the desired agronomic and genetic traits from which high yield and resistant sources could be developed (Altieri, 1991; McNeely *et al.*, 1990). and have full potentials in the wake of climate change events such as drought etc.. Unfortunately, these traditional crops only attain special status in times of natural calamity and famine. Yet these are staple crops that farmers are already very familiar with, offering several ideal qualities as crops for food security in the fragile Himalayan Mountains.

Nutritive value of traditional crops

The main nutritional value of traditional crops like finger millet, hog millet and barnyard millet, lies in their potential ability to provide one of the cheapest sources of dietary energy, in the form of proteins and carbohydrates, in the Himalaya. The majority of traditional crops of the mountain have the highest calorific yield. Such traditional crops are particularly valuable in the mountains where most of the population depends on vegetable protein and carbohydrate foods as dietary staples. Based on the study conducted by Maikhurie *et al.*, 2003, it is quite evident that very few people in the Himalayan villages suffer from a simple protein deficiency. Increasing the consumption of traditional food items could provide the much needed protein currently provided by other foods such as cereals and pulses. Traditionally, in the Himalayas many of these traditional crops supplement the wheat and rice meal. The significance of these crop resources cannot be assessed from the number of hectares under cultivation and the amount of production (which cannot be large) but from the kind of contribution they make to support lives under difficult Himalayan conditions. For example, in areas where meat and milk were historically scarce the domestication of grains with superior protein content was considered to keep the population well nourished. While in the Andes, crop like quinoa improved the daily diet of the native people, so in the Himalayan region, crops like amaranth, chenopods, hog millet, buckwheat and several kinds of beans and little known wild edibles helped nourish the people (Pratap,1990; Maikhuri *et al.,* 1991 b, c).

Study area and climate

Uttarakhand (28°43′ and 31°28′ N latitude and 77°49′ and 81°03′ E longitude) is a hill state located in the central part of the India and shares an international boundary with China in the north and with Nepal in the east. The state encompasses an area of 53,483 sq. km, which accounts for nearly 15.5 % of the total geographical area of western Himalaya and 1.63% of the total land area of India. The total population of the state is over 10 million (census-2011). Vegetation varies according to altitude and climatic conditions, from tropical deciduous forest in the foothills to Himalayan temperate forest at middle altitudes, coniferous, sub-alpine and alpine forest at higher altitudes, giving way to alpine grasslands and high altitude meadows. The region can be divided into three markedly different agro-climatic zones along the elevational gradient (vertical zonation) viz., lower altitude, 500 to 1000 m, middle altitude between 1000 to 1800 m and higher altitude, above 1800 m. According to traditionally accepted criteria agricultural land in the region is identified either as the rainfed (locally known as *Ukhar*) or the irrigated (known as *sera*). Of the two, the former is predominant form of land use since only about 15% land falls in the latter category.

Methodology

A reconnaissance survey was carried out in 48 villages of central Himalayan region to collect the information regarding the perception of local people about climate change impact over a decade in the agriculture sector through application of different PRA tools. These villages were selected randomly along an altitudinal gradient. Structured and semi-structured questionnaires were developed and discussed with different aged people. Before collecting the actual data, the survey and questionnaires were tested and improved in order to record information obtained from people properly. The adaptation and mitigation strategies developed by the local community related to climate change was assessed through field surveys by undertaking formal and informal group discussions and personal interviews and further data verification by actual field checks. Face to face interviews with individuals (both men and women) of different age categories was also carried out to identify the past and current coping strategies developed by tribal, non-tribal and traditional and indigenous communities to adapt and mitigate global environmental changes or impact of climate change. The interpretations about climate change and their ecological and socio-economic implications were drawn based on the methodology applied by Saxena *et al.*, 2004; Maikhuri *et al.*, 2003, 2009, related to climate change predictions/ projections, interaction of climate changes with other global change such as land uses, land cover change, biological invasion, economic globalization and socio-cultural transformation. Rapid participatory survey in selected villages in several valleys of central Himalaya across elevational gradients were carried out between 2000- June 2010 to assess the people's perception related to climate change and its interactions with other changes and impacts on ecosystem and livelihood. Identification of strategic actions to cope up with and adapt to changing biophysical and socio-economic conditions as revealed by the people also noted. During the survey conducted by a team of researchers, various tools such as focused group discussion, questionnaire method and recall methods were used to elicit the information on impact of coping mechanisms, adaptation/ mitigation or other changes to climate change (Makhuri *et al.*, 2003, 2007, 2009).

Result and discussion

Climate change and mountain agrobiodiversity

Cropping patterns in Himalaya are built around two seasons that are monsoon/ rainy season and the winter season. With a belief of occurrence of absolute crop failure due to bad climate only in one cropping season in a year, farmers tend to stock staple food required for a period of 6 months. Over the last 50years, farmers could recall complete failure of both crops only in 1966-67 in a few high elevation villages. People coped such a situation by consuming wild staple food like dried fruits of *Pyrus pashia*, and *Aesculus indica* and by selling non- timber forest products and by earning wages in far off places. However, policy of supplying a quota of staple food at subsidized price since 1970s have drawn farmer's attention away from local production based food security system (Semwal *et al.*, 2004; Singh *et al.*, 2008). Even though sowing of winter crops has been delayed by about a month due to late winter precipitation in recent years, harvesting time seem to be unchanged possibly because of warming.

Farmers view two major risks to crops, the risks arising from (a) uncertainty of monsoon rainfall and (b) cultivating distant fields that demand huge labor and time spent in travel/transport. Maintenance of a heterogenous village landscape is a reflection of indigenous ways of risk management. The rainfed agroecosystem is characterized by both climate and distance related risks, rainfed agroforestry system by only climatic risks, the irrigated crop system by only distance related risk and the homegarden system by neither of the two risks (Singh *et al.*, 2008). Within an agroecosystem type, farmers cope climatic risks by growing crops/ cultivars based on indigenous knowledge on their responses to different ecological conditions (Bardsley, 2003). After the first monsoon showers in April/ May, Muatha and Bhagan cultivars of barnyard millet and Jhalarya and Chauras cultivars of fingermillet are sown in a few fields. If monsoon commences by mid –June, maize is sown in about 50% of the remaining fields. If maize growth is normal till 20-25 days after sowing, soybean is intercropped between maize rows and Mungerikuad cultivars of finger millet is sown in the remaining vacant fields. If the crop growth is poor during first month after sowing, maize field are ploughed afresh and sown with Mungerikuad cultivar of fingermillet. If monsoon does not commence by mid-June, all cultivars of barnyard millet and fingermillet are sown, covering almost equal areas. Jhaidu and Khimanand ki Ghodi are local rice cultivars able to withstand hail storms while Rekher or Syal Satti and Misri and Thanga varieties of wheat can survive heavy snow fall occurs during early stages of crop growth (Saxena and Rao, 2009).

Farmers classify crops in three groups based on their responses to various stresses/risks together with their economic values (a) economically more-valuable crops with poor performance under extreme rainfall regimes, low soil fertility levels and weed abundance including maize, soybean, paddy, wheat, lentil, potato,, buckwheat, amaranths and green vegetable, (b) economically less-valuable crops with ability to perform under unfavorable climatic conditions, low soil fertility levels and weed abundance including fingermillet, barnyard millet and barley and (c) economically more-valuable crops with ability to perform under unfavorable climatic conditions, soil stresses and weed abundance including sesame, cowpea, black pea (*Pisum arvense*), horsegram and pigeon pea. Though many of farmer's perceptions about crop-environment relations are substantiated by scientific evidences (Maikhuri *et al.*, 1996; Singh *et al.*, 2008), there is a need of undertaking long term comprehensive programs for validation and enhancement of local knowledge system. A traditional crop like *Panicum miliaceum* is unique in that its maturity time is as short as less than two months but is not viewed as a delicious food. Increasing the area under this crop following bad climate seasons is a common way of achieving food sufficiency. Farmers apply higher levels of inputs available in limited quantities to the perceived low risk agro ecosystems as compared to the more risky ones (Carter and Murwira, 1995).

Conversion of rainfed to irrigated farming reduces the risks of climatic uncertainty and improves productivity (Bhatnagar *et al.*, 1996; Maikhuri *et al.*,

1997) but has not progressed much for two reasons. First, farmers face a shortage of manure (due to scarcity of forest resources) required for performance of irrigated crops and are unable to afford the chemical fertilizers. Second, labour productivity from irrigated crop system is lower than other traditional land uses like home gardens/kitchen garden. Highly productive indigenous irrigated farming systems do exist in situations where rainfed crops fail to survive (e.g. cold desert) or when population pressure exceeds the carrying capacity of rainfed agriculture or when farmers do not have any source of income other than irrigated crops (Rao and Saxena, 1994. Chandrasekhar *et al.*, 2007).

A trend of replacement of traditional staple food/fodder crops by cash crops is progressing fast since last couple of decade partly because of a socio-cultural transformation from subsistence to market economy often supported by development programmes enabling availability of food and modern agricultural inputs at subsidized rates of government supported marketing system (Singh *et al.*, 1997; Semwal *et al.*, 2004). Agro-ecosystem in the Himalaya heavily dependent on forests for manure and livestock feed. The ongoing changes in cropping patterns are such that manure input rates have increased along with an increase in soil erosion rates but fodder production from farmland has decreased, which in turn has increased threats to forest ecosystem functions as a result of higher intensities of litter removal and grazing in forests (Sen *et al.*, 1997; Maikhuri *et al.*, 2000). The tendency for maximization of income has marginalized the traditional values attached to crop-environment compatibility, social integrity achieved through exchange of complementary crop products and farmer's selection/conservation of crop genetic diversity. Over time, many local communities have gained some understanding about the risks and uncertainties and have been found to grow cash crops to an extent that their traditional food security system is least disturbed (Maikhuri *et al.*, 2000; Semwal *et al.*, 2004; Saxena and Rao, 2009).

As biological potential of traditional crops/cultivars is poorly understood, their loss is not in the interest of the scientific and wider community too. Traditional farmers has selected and conserved a wide variety of crops and cultivars to cope up with the risks, uncertainties and extremes of monsoon and huge micro-edaphic variability in the mountains. Pests and diseases are common in warm-cold regions but not in extreme cold arid region, indicating a possibility of higher risks of damaging agents in warmer climate. Farmer selected cultivars/crops may reduce such future risks.

Farmers' perceptions on climate change impact on agriculture sector

Many traditional communities have responded to changing environments (Grove, 1996). Analysis of indigenous knowledge could provide insights on changing climate and its impacts. Deductions from people's perceptions, however, will be limited to a time scale, which is within the range of human memory. Farmers may hide or provide inaccurate information and hence cross-checking of their perceptions are warranted (Sen *et al.*, 2002). People's perceptions derive not from any direct measurements of climate but from the way climate affects their immediate surroundings and livelihood. For people in central Himalaya, a good climate meant: sporadic low rainfall during March, mid-May, peak rainfall during July-August, moderate rainfall/heavy snowfall during December/January and the absence of cloud burst events. People consider onset of monsoon to be more uncertain compared to other phases of rainfall. Climate changes felt in the recent decades included a shift in peak rainfall time from July/August to August/ September and winter precipitation from December/January to January/February, increase in frequency of cloud-burst and warming (Table 1).

Table 1. People's perceptions on climate change in central Himalayas (Saxena *et al.*, 2004; Maikhuri *et al.*, 2003).

Kind of change	Evidence
Warming	Decline in snowfall period, depth and persistence, decline in apple yield, success of cabbage/pea/tomato cultivation in high elevations in recent years, shortening of maturity period of winter crops, increased pest infestation.
Decline in rainfall during March – May	Large scale mortality, abandonment of *Panicum milliaceum* in rainfed area, declining yields of Amaranthus.
High rainfall during August/ September instead of the normal peak in July/August	Damage to rainy season crops when they are close to maturity, increased frequency and severity of landslides
Winter precipitation in January/February instead of December/January and decline in intensity of snowfall	Delayed sowing of winter crops, decline in barley and wheat yields
Increase in instances of cloud burst	Heavy losses of life and property

Indicators of climate-change impacts were analyzed through interactive discussions with 350 respondents in two different age groups (i.e., AG1, 20-50 years and AG2, 50-80 years or above). The majority of respondents in both age groups agreed that during recent decades there had been many changes observed in the climate and they cited various examples. It was observed that respondents who were 50-80 years or above had a deeper understanding, keener observation, more in-depth knowledge, and more experience of the changes in local climate changes than the respondents who were 20-50 years of age (Fig. 1). In-depth field observations of a team of scientists from GB Pant Institute of Himalayan Environment and Development (GBPIHED) (Garhwal Unit) over the past 18 years and local people's perceptions revealed that advancement of flowering, leafing, and fruiting time (15-20 days) of medicinal and aromatic plants (i.e., *Rhododendron arboretum, Allium stracheyi, A. humile, Betula utilis, Meconopsis aculeate* and *Saussurea obvallata*) and some prominent wild edible species (i.e., *Rebis orientale, Rosa webbiana*,and *R. sericea*) has been noticed and they considered it the most striking evidence of climate change (Fig. 1).

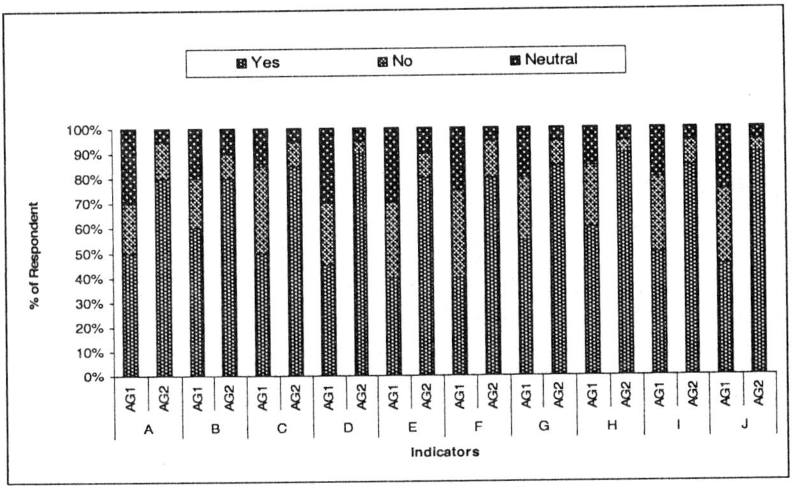

Figure 1. Farmer's perceptions about the various indicators of climate change in central Himalaya (% of farmers in different age groups responding in each indicator category) (Source: Maikhuri *et al.*, 2009)

A. has the climate changed?, B. Increase in temperature, C. Decrease in snowfall, D. Decrease in rainfall/shift in rainfall, E. Shift in crop maturation, F. Shift in crop showing calendar, G. Increase in pest/disease in crops, H. Change in agricultural land use, I. Decrease in water availability in irrigated land, J. Change in agro-biodiversity.

*Total no. of respondents 650 (300 respondents of 50-80 years or above and 350 respondents of 20-50 years) belonging to central Himalayan mountain region.

More than 50% of respondents in the 20-50 age group and 90% of the respondents in the 50-80 or above age group noticed the significant decrease in snowfall and cover in the high altitude villages i.e. Malari, Gamsali, Niti, and Dunagiri located between 3,000-3,600 masl because of their recollections of 30 years or more. The recession of the Satopanth, Dunagiri, and other glaciers located in the catchment is another important indicator of climate change in the region according to local people involved in trekking and mountaineering activities. The most significant piece of evidence is a decrease in water resources used by livestock, particularly in several alpine pastures, low-altitude forests, and grazing areas over a period of 10-15 years according to transhumant pastoralist communities. They attributed this change to a gradual increase in temperature and the consequent drying up of water bodies. Nonetheless, the possibility of modification of changes caused by the altered climate by non-climatic factors cannot be ruled out (Maikhuri *et al.* 2003).

Vulnerability to climate change

Impact of drought on hill/mountain agriculture

Agriculture in central Himalayan region is highly dependent on weather and seasonal rain and changes in global/regional local climate have a major effect on crop yield and food security in this region. Agriculture coupled with horticulture is an important source of sustenance of the traditional societies. Irregular rainfall, delayed rainfall, low rainfall and no rainfall, snowfall, change in climatic condition and rising temperatures reduced yield of agricultural crops, vegetables and horticultural crops in the region. Changes in rainfall patterns observed in this region have seriously affected agriculture, which already had to cope with highly irregular and unpredictable rainfall patterns. Decreases in rainfall had profound repercussion on the flows of the thousands of small streams that provide an essential source of water to irrigated land on which livelihood of the people dependent to large extent. Declines in river flows and runoff in those regions have typically been more pronounced than those in rainfall. However, decreased in rainfall intensity over several years (from 1998-2007) have contributed to the shrinking of rivulets, springs (Kumar *et al.*, 2008). However, extreme weather events have also become more common. Increasing temperatures and delayed or changed rainfall patterns will affect agriculture dramatically, by changing the geographical distribution of areas suited to the different crops. As a result, people of this region will have to adjust with the practices of agriculture as they do and cropping pattern as they grow with changing climatic conditions. Besides crops such as traditional legumes (*Vigna unguiculata, Vigna angularis*) important summer legumes growing between 1,000-2,000 masl are facing problems of fruit setting because of the shift in peak rainfall time and other climatic factors revealed by the farmers (Maikhuri *et al.*, 2009, 2010).

Traditional farming practices are now threatened due to increased temperatures, reduced water resources and drastic changes to the frequency and volume of precipitation patterns in this region, typically resulting in either extended periods of drought. *Hordeum himalayens* an under utilized crop of huge economic importance in central Himalaya, has been showing significant reduction in its cultivated area over the years (Maikhuri *et al.*, 1996, 1997, 2001). The effect of temperature on photosynthetic characteristic of two species of Hordeum (*H. himalayens*) grown in high altitude between 2200-3000 m asl and *H. vulgare* cultivated at low altitude (300-1500 m asl) reveals that *H. himalayens* is more sensitive to photosynthesis and higher temperature than the later one and it could be one of the critical factors responsible for low yield and that has resulted to reduce the area under its cultivation (Joshi and Palni, 2005). Changes in rainfall pattern has been observed during last more than six years that has resulted increase in pests, diseases and invasive species and has direct impact on crop productivity and food security of the people of the central Himalaya (Maikhuri *et al.*, 2009, 2010). Recently, Himalayan region experienced delayed monsoon and because of this there was no rain during the period farmers normally transplant wet rice crops (Table 2). Consequently, the thousand of hectare of land which was earlier under irrigated conditions for growing rice and wheat was left either barren or farmers switched towards rainfed agriculture and it has drastically reduced the crop productivity (Table 3). In the central Himalaya changing rainfall patterns and difficulties in predicting the weather have made increasingly difficult for farmers to plan their crop sowing/planting in rainfed and irrigated lands. Whereas the rains used to be expected usually in third week of

June, the past decades showed greater variability in rainfall patterns (Kumar *et al.*, 2008). This shift has become more marked in recent years, with rainfall as late as August than or as late as September (very intensive rains lasted at least for two weeks) which damaged the crops such as Potato and Rajma, Amaranth and other pulses and reduced the yield upto 70-80% in 2010 (farmers as well as personal observations).

Table 2. Impact of climate change, vulnerabilities and adaptation measurements in Central Himalaya (Source: Saxena *et al.*, 2004; Maikhuri *et al.*, 2003; Saxena and Rao, 2010)

Kind of change	Vulnerable sectors	Adaptation/ mitigation measures
Decline in snowfall/ rainfall	• Decline in snow-covered area around clearly visible peaks • Decrease in depth and persistence of snow around settlements in higher altitudes • Decline in apple yield as it needs proper chilling during winters for proper fruit yield.	• Decline in transhumant practices as poor winter precipitation is believed to follow poor fodder quality as well as quantity. • Replacement of apple by annual cash crops like pea, tomato, cauliflower, chilly and cabbage under rainfed conditions in higher altitudes and irrigated conditions in lower altitude
Decline in rainfall during March-May	• Adverse impact in terms of decline in yield of Kharif crops due to large scale mortality and/or poor growth in the initial stage of crop growth.	• Abandonement of crop, e.g., *Panicum miliaceum* which used to be sown in March. This crop matures over a period of 3 months. The crop is badly affected if rainfall is delayed. • Casual management of traditional staple food crops grown during rainy season. • Replacement of Amaranthus paniculatus by cauliflower, cabbage and potato, search of off-farm employment opportunities. • Innovation of new agricultural practices characterized by high labour productivity and stress tolerant cash crops e.g. bush fallow agriculture.
High rainfall during August/ September instead of the normal peak in July/ August	• Damage to rainy season crop when they are close to maturity. • Huge landslides blocking roads during Agust-September	• Search for off-farm opportunities
Winter precipi-tation in January/February instead of December/ January and	• Delayed ploughing/sowing of wheat, barely, naked barley and mustard earlier it used to be done in November but now in December.	• Replacement of traditional cultivars of wheat by high yielding varieties. • Replacement of barely by green pea.

decline in intensity of snowfall	• High yielding varieties of wheat and green pea which did not perform well earlier can perform well at higher altitude. • Decline in barley and wheat yield.	
Increase in instances of cloud burst	• Heavy losses of life and property due to heavy downpour over a short period of time	• Reestablishment of dwellings

Table 3. Decreasing area (ha) under rice cultivation due to climate change in central Himalayan mountain region (Source: Amar Ujala, 12 June, 2010).

Year	Area (ha)
2005	108048
2006	107365
2007	107161
2008	106622
2009	92192
2010	48223

Impact of insect and pest incidence on agriculture

Climate change could have positive, negative or no impact on individual plant diseases, but with increased temperatures and humidity many pathogens are predicted to increase in severity. In central Himalayan region, conditions are usually more favourable for the proliferation of insect pests in warmer climates due to climate variability. Longer growing seasons may enable a number of insect pest species to complete a greater number of reproductive cycles during the spring, summer, and autumn. Warmer winter temperatures may also allow larvae to winter-over in areas where they are now limited by cold, thus causing greater infestation during the following crop season. Most studies have concluded that insect pests will generally become more abundant as temperatures increase, through a number of inter-related processes, including range extensions and phenological changes, as well as increased rates of population development, growth, migration and over-wintering (Altieri and Nicholls 2004). The most likely consequences are shifts in the geographical distribution of host and pathogen and increased crop losses, caused in part by changes in the efficacy of control strategies. Altered wind patterns may change the spread of bacteria and fungi that are the agents of wind-borne plant diseases for instances *Puccinia graminis* (Saxena and Rao, 2009; Coakley *et al.*, 1999).

During the recent past, as a result of rapid changes in land use caused by socio-cultural and economic changes and various environmental perturbations the biodiversity of agricultural crops in the whole of region has changed steadily (Maikhuri *et al.*, 2001). However, in recent decades, climate change has emerged as another threat to agrodiversity.

Notwithstanding, reduction in winter snowfall and spring rainfall and melt-water flows will produce a deficit of soil moisture that could limit any increase in yields resulting from temperature increases. It is believed that even minor changes in temperature could have a major impact on the severity of diseases. *Amaranthus* crops are most vulnerable to climate change as observed during the recent past when this crop was severely attacked by a disease called *Hymenia rickervalis* between 1,000 – 1,800 masl, whereas between 2,200 – 2,800 masl it grew well. It is assumed that poor rainfall during July and high temperatures and humidity, particularly in first and second weeks of September provided favourable conditions for the moth that damaged the crop and reduced the crop yield (Maikhuri *et al.*, 2009, 2010). Diseases such as rust and blight were common in cereals and potato crops, and legumes such as *Phaseolus* spp were infected through soil- borne insects such as *Coleoptera* species (Maikhuri *et al.*, These insects damage the crops in the early stages of seed germination. One of the reasons for the occurrence of disease in these crops could be that the climatic conditions are favorable for the life cycles of the insects, i.e., an increase in moisture or humidity or milder winters (between 500-1,500 masl) in the lower regions. High-altitude agriculture in this region is definitely in transition and a rise in temperature in future may enhance agricultural productivity (Table 2). Some studies (Rosenzweig *et al.,* 2001; FAO, 2004c) agree that higher temperatures and longer growing seasons could result in increased pest populations in temperate regions of Himalaya. CO_2 enrichment and changes in temperature may also affect ecology, the evolution of weed species over time and the competitiveness of C3 v. C4 weed species (Ziska, 2003). Warmer winter temperatures would reduce winter kill, favouring the increase of insect populations. Overall temperature increases may influence crop pathogen interactions by speeding up pathogen growth rates which increases reproductive generations per crop cycle, by decreasing pathogen mortality due to warmer winter temperatures, and by making the crop more vulnerable. Overall the high altitude agriculture in this region is in transition, and a rise in temperature in future, under certain conditions, may even enhance agricultural productivity.

Adaptation and coping strategies

Agricultural production has a great deal to contribute to the short and long term reduction of poverty. Careful analysis and evaluation are required so that agriculture and related developmental strategies can be reoriented towards better use of local resources so as to contribute more effectively to improving the livelihood of the rural poor and ensuring sustainable agriculture together with conservation of genetic diversity of crops, cultivars, agro-ecosystems and landscapes so as to provide greater variety of adaptations to changing climatic conditions (Maikhuri *et al.,* 2001b).

Improvement in traditional soil fertility management

As agriculture is dependent on forests for manure and fodder, reduction in intensity of biomass removal from forests without any threat to agroecosystem functions is crucial for forest conservation. Application of oak residue based manure enables crop yields 15% higher compared to pine residue based manure partly because of higher rates of nitrogen mineralization coupled with better synchronization of nutrient release and crop uptake in the former (Rao *et al.*, 2003).

Water management

Global warming will aggravate water stress, a factor often limiting crop yields and life quality. The traditional systems centered on minimum inputs for water purification, storage and canalization, minimal interference with natural hydrological processes and minimal risks of damages likely from high rainfall events were by and large abandoned when water supply was treated as a government service to the people. With experiences of the large scale failure of the new system over the last few decades (Rao and Saxena, 1994; Maikhuri *et al.*, 1997, 1998; Rao *et al.*, 1999), innovations in water technology and management sectors that can be sustained in the likely global warming scenarios are needed.

Enhancing adoptive capacity in traditional agroforestry through tree management

A successful and well managed integration of multipurpose trees agriculture land inevitably results in diversified and sustainable crop production, in addition to providing a wide range of environmental benefits. Scattered agroforestry trees are distinguishing features of settled upland farming. Lopping is a tool to regulate tree-crop competition for optimizing multiple benefits from the system. Farmers usually lop all branches during winter season when accesses to as well as availability of fodder/fuelwood from forests are constrained by harsh climate. Semwal *et al.* (2002) have shown that retention of 25% of branches together with increase in tree density in private farmland will improve tree vigour and ecological functions without any decline in crop yields. Research is needed to identify interventions that lead to agricultural sustainability such that pressure on forests is reduced. As adaptation is yet developing as science, the role of agroforestry in reducing the vulnerability of agricultural systems and the rural communities that depend on them for their livelihood to climate change or climate variability should be more strongly emphasized.

Reconcile agrobiodiversity conservation and development

In spite of the many virtues of traditional crops, precious genetic diversity, the rivet of ecosystem stability, is gradually being lost (Maikhuri *et al.*, 2002). As biological potential of traditional crops/cultivars is poorly understood, their loss is not in the interest of the wider community. A realization of negative consequences of high yielding varieties, viz., dependency on external agencies for seeds, fertilizers, irrigation and pesticides, drastic yield losses under unfavorable climatic conditions and low input management and lower fodder production compared to traditional varieties, in recent years has rejuvenated local efforts towards agrobiodiversity conservation in Hanval valley of Tehri Garhwal. *In situ* conservation of traditional crops and cultivars could succeed when these crops are strongly linked with the economic development of hill farmers. Pragmatic multidisciplinary research efforts are needed to evolve farming systems with appropriate selection of crops in view of future climate change which can provide enough quality food and economic security to the people of the region together with conservation of the traditional crop wealth, sustainability of the production systems and environmental conservation. Maintaining and enhancing traditional agrobiodiversity and cultural diversity in this part of Himalaya which offer significant adaptation potential to respond to the challenge of climate change.

Promotion of organic cultivation emphasizing traditional crops

The agriculture practiced in the hills/mountains is also known as low external input agriculture. As the steep slopes lead to high percolation losses of water, farmers, have traditionally overcome this by adding large quantities of farmyard manure (FYM) to the fields, more for its contribution to enhance soil fertility. The application of chemical fertilizers, pesticides, insecticides, herbicides/ weedicides etc. very least in the mountains. Since the agriculture practiced in the hills is organic, therefore, there is a need to look at sustainability for both agricultural production as well as food security. To give further impetus to organic agriculture as well as to locally based food security, government needs to widen the food basket to include locally grown traditionally and nutritionally rich organic food crops. Ecological and organic farming mitigate climate change by reducing greenhouse gas emissions and increasing carbon sequestration in plants and soil (Maikhuri *et al.*, 2002).

Bioprospecting and value addition of traditional mountain crops of nutraceutical potential

Promoting a supportive environment for the development of high quality products and services from this region as a means to improve livelihoods and protect environments and encourage more active involvement by the local communities and private sector in all parts of the value chain for high quality value added Himalayan agro-food products.

Capacity building and skill development in the field of eco and climate friendly technologies

Technologies change is an important instrument in the continuous process of socio-economic development but due to poor access to appropriate technologies is one of the main cause of poverty, drudgery and natural resource degradation in the Himalaya (Maikhuri *et al.*, 2010; Rawat *et al.*, 2010). Therefore, establishment of rural technology demonstration and training centre in rural set up and capacity building/skill development of farmers could provide viable option for improving yield potential of farm produce, income generation from off- farm activities and thus reduce the existing pressure on forests and natural resources (Maikhuri *et al.*, 2007a, 2010). The appropriate technologies suitable for high altitude region such as protected cultivation, organic compost and bio-fertilizers, off-farm technologies and other supporting technologies would substantially increase the yields on marginal land and would also provide suitable mitigation and adaptation potential to face the challenges of climate change.

Institutional cooperation, coordination, collaboration and capacity building to address climate change in various sectors

In central Himalayan region particularly Uttarakhand state has many Research and Development institutions with significant infrastructure and scientific/technical capacity. However, so far these institutions have not much focused on climate change research, which includes modeling, field ecological and biodiversity related studies etc. There is inadequate capacity in several research and developmental institutions working on environmental and conservation issues in relation to climate change. There is an utmost need to create awareness and enhance capacities at individual and institutional level and also include other stakeholders i.e. NGOs etc. Local communities depending on natural resources for their livelihood have poor financial, technical and institutional capacity to adapt to adverse impacts of climate change. Therefore, it is important to enhance capacities of local people's (whose survival entirely depends on marginal agricultural land and their surrounding forest vegetation) who are likely to be vulnerable to projected climate impacts.

Conclusion

The suffering of people inhabited in hills/mountains is gradually increasing and their food and nutritional security and cultural identity is declining because they have been neglected at both policy and practice levels by government. The lack of understanding and appreciation of the micro-macro linkages in the context of global changes will have a profound impact on the way business of research and development is conducted in the hilly part of the Central Himalayan (Uttarakhand). Whether it is the opening of global markets or emerging WTO provisions, there will be serious implications on the livelihood security of the marginal farmer, especially in the hills where people survive under vulnerable environments. Agriculture in the hills of central Himalaya is definitely in transition (Maikhuri *et al.*, 2009, Farooquee and Maikhuri, 2010). The continuing change in climate would have a major impact on local agriculture and thereby affect the food supply at local and regional level. In mountain particularly in high altitudinal region of central Himalaya, the productivity of crops is adversely affected by low temperature and short growing seasons. In high altitude areas, the global shift in the environment is leading to rise in temperature and in very near future, however, it will bring more opportunities with the changing environment particularly for cash crops like amaranth, kidney beans, wheat and vegetables like (redish, tomatoes, cabbage, chillies, peas) and medicinal plants if water and soil are adequate (Maikhuri *et al.,* 2003). In addition, horticultural crops and trees those provide timber may produce higher yield. A long growing period and accelerated soil decomposition, leading a better nutrient intake by trees and other plants as well as to higher biomass and productivity, may be added advantage and positive impact of the climate change. Therefore, there is a need to follow the right policies, implement strong institutional initiatives, research and to incorporate some positive options which can provide food and economic security to the people together with conservation of traditional crops and environmental management in the wake of climate change impact.

Traditional ecological knowledge as well as people experiences about the pattern of climate change and its impact on forest, agriculture, livestock and human health to be analyzed in an integrated manner so as to provide possible indicator of change and adaptation mechanism to face the challenges of climate change. The cultural and ethnic diversity and traditional knowledge systems available in the region to be given priority in adapting climatic change and be recognized and enhanced through appropriate policy and investment. Adoption of large scale bio-resource conservation and utilization strategies and better integration of agriculture, livestock, forestry and other local food processing in order to help diversify livelihood activities to make local food systems more resilient to climate change. Interface with policy issues, administration, local communities and research and academic institutions regarding the broad aspects of adaptation options and livelihood improvement to be incorporated in R&D agenda. Capacity building and skill development of all stakeholders in climate change processes and trends, mitigation and adaptation measures and food security options in central Himalayan region along with demonstration and implementation of climate and eco-friendly technologies to be given top priority in near future. There is a need to develop climate proof and pest control genotypes for biotic (diseases, insects etc) and abiotic (drought, flood, heat, cold, salinity) stress management either by traditional plant breeding, or genetic modification. Research and development solutions required to solve problems need to be location-specific and matched to the huge agroecological/climatic diversity of the central Himalayan region.

Acknowledgements

Authors are thankful to Director, G.B. Pant Institute of Himalayan Environment and Development, Kosi-Katarmal, Almora for providing facilities and to the TSBF/GEF/UNEP for financial support.

References

Altieri M A and Merrick L C (1987), *In situ* conservation of crop genetic resources through maintenance of traditional farming systems. *Economic Botany* , 41: 96.

Altieri M A and C I Nicholls (2004), Biodiversity and Pest Management in Agro-ecosystems. 2nd edition. Haworth Press, New York.

Arora R K (1980), Native food plants of the north-eastern tribals. In Jain, S .K. (ed.) *Glimpses of Indian Ethnobotany*, pp. 91-106 (New Delhi: Oxford IBH Publishing Co.).

Arora R K and Nayar E R (1984), Wild relatives of crop plants in India, 97 pp. NBPGR *Science Monograph*, 7

Barddsley D (2003), Risk allevation via in situ agrobiodiversity conservation: drawing from experience in Switzerland, Turkey and Nepal. Agriculture, Ecosystem and Environment, 99: 149-157.

Bhatnagar P R, Srivastava R C and Bhatnagar V K (1996), Mangement of runoff in small tanks for transplanted rice production in the mid-hills of northwest Himalaya. *Agriculture and water management*, 30: 107-118.

Carter S E and Murwira H K (1995), Spatial variability in soil fertility managment and crop response in Mutoko commmunal area, Zimbabwe, *Ambio*, 24: 77-84.

Chandrasekhar K, Rao K S, Maikhuri R K and Saxena K G (2007), Ecological implications of traditional livestock husbandry and associated land use practices: a case study from the trans- Himalaya, India. Journal of Arid Environments, 69: 299-314.

Coakley S M, H Scherm and S Chakraborty (1999), Climate Change and Disease Management. Ann. Rev. Phyto. 37:399-426.

FAO (Food andAgriculture Organization), (2004c), Data Base. Food and Agriculture Organization of the United Nations, Rome.

Farooquee, Nehal A and R K Maikhuri (2009), Communities and their agrobiodiversity: Priorities for agriculture in Uttarakhand Himalaya, India. *Outlook on Agriculture* 38(4): 383-389.

GoI (Government of India) 2004, Forest and Wildlife Statistics: India 2004.

Grove J (1996), The century time scale. In: Driver, T.S. and Chapman, G.P. (eds.), Time-scales and Environmental Change. Routledge, London. Pp. 39-87.

IPCC (2007), Working Group II contribution to the fourth assessment report, Climate Change: Climate change impacts, adaptation and vulnerability (Geneva: WHO and UNEP IPCC, 2007).

Kireet Kumar, Neh Joshi and Varun Joshi (2008), Climate variability vulnerability and copping mechanism in Alaknanda Catchment, Cnetral Himalaya, India, Ambio, 37 (4 J): 286-291.

Maikhuri R K (1991b), Nutritional value of some lesser known wild food plants and their role in tribal nutrition. A case study from northeastern India *Tropical Science*, 31: 397-405.

Maikhuri R K, R L Semwal, K S Rao, S Nautiyal and K G Saxena (1997), Eroding traditional crop diversity imperils the sustainability of agricultural systems in Central Himalaya, *Current Science*, 73: 777-782.

Maikhuri R K, L S Rawat, Vikram Negi and V K Purohit (2007a), Eco-friendly appropriate technologies for sustainable development of rural ecosystems in Central Himalaya. G.B.Pant Institute of Himalayan Environment and Development, 45p.

Maikhuri R K, Nautiyal M C and Khali M P (1991c), Lesser known crops of food value in Garhwal Himalaya and strategy to conserve them. FAO/IBPGR Plant Genetic Resources Newsletter, 86: 33-66.

Maikhuri R K, Nautiyal S, Rao K S, Chandrasekhar K, Gavali R and Saxena K G (2002), Analysis and resolution of protected area-people conflicts in Nanda Devi Biosphere Reserve. *Environmental Conservation*, **27:** 43-53.

Maikhuri, R. K. Nautiyal, S, Rao, K. S , Chandrasekhar, K, Govall, R and Saxena, K. G. (2000), Analysis and resolution of protected area- people conflicts in Nanda Devi Biosphere Reserve, India, *Environmental Conservation* **27:** 43-53.

Maikhuri R K, Rao K S and Saxena K G (1996), Traditional crop diversity for sustainable development of central Himalayan agroecosystems, *International Journal of Sustainable Development and World Ecology*, 3: 8-31.

Maikhuri R K, Semwal R L, Rao K S, Saxena K G and Das A K (2001b), Indigenous techniques of agricultural soil fertility maintenance in the central Himalaya, Ecology, Enviornment and Conservation 7: 15-20.

Maikhuri R K, Semwal R L, Rao K S and Saxena K G (1997), Rehabilitation of degraded community lands for sustainable development in Himalaya: a case study in Garhwal Himalaya. *International Journal of Sustainable Development and World Ecology*, 4: 192-203.

Maikhuri R K, U Rana, K S Rao, S Nautiyal and K G Saxena (2001), Promoting eco-tourism in the buffer zone areas of Nanda Devi Biosphere Reserve: An option to resolve people- policy conflict. International Journal of Sustainable Development and World Ecology **7:** 333-343.

Maikhuri R K and K S Rao (2002), Inventory, commercialization and conservation of agrobiodiversity for sustainable development of the buffer zone villages of Nanda Devi Biosphere Reserve. In: Sharma, J.K., P.S. Easa, C. Mohanan, N. Sasidharan and R.K. Rai (eds.), *Proceedings of Biosphere Reserve in India and Their Management*, pp. 128-130.

Maikhuri R K, K S Rao and R L Semwal (2001), Changing scenario of Himalayan agroecosystems: loss of agrobiodiversity, an indicator of environ-mental change in Central Himalaya, India. *The Environmentalist* 20: 23-39.

Maikhuri R K, L S Rawat, Vikram S Negi, Prakash Phondani, Abhay Bahuguna, K P Chamoli and N A Farooquee (2009), Impact of climate change and coping strategies in Nanda Devi Biosphere Reserve, Centaral Himalayas, India. In: *Proceedings of the International Mountain Biodiversity Conference*. ICIMOD, Kathmandu, Nepal, pp.138-146.

Maikhuri R K, Rao K S, Patnaik S, Saxena K G and Ramakrishnan P S (2003), 'Assessment of vulnerability of forest, meadows and mountain ecosystems due to climate change.' *ENVIS Bulletin* 11(2): 1-9.

McNeely J A, Miller K R, Reid W V, Mittermeier R A and Werner T B (1990), Conserving the world's biological diversity, 193 pp. (Gland, Switzerland: IUCN, WRI, CI, WWF and World Bank).

Nautiyal S, K S Rao, R K Maikhuri and K G Saxena (2002), Transhumant patoralism and sustainable development: A case study in the buffer zone of the NDBR, India, *Mountain Research and Development* **23 (3):** 255-262.

Pratap T (1990), Biological diversity and genetic resources: Some issues for sustainable mountain agriculture. Paper presented in the International symposium on *Strategies for the Sustainable Mountain Agriculture*, (Kathmandu, Nepal: ICIMOD).

Rao K S and Saxena K G (1994), Sustainable Development and Rehabilitation of Degraded Village Common Lands in Himalayas Publication No. 8, Bishen Singh Mahendra Pal Singh, Dehradun, India.

Rao K S, Semwal R L, Maikhuri R K, Nautiyal S, Sen K K, Singh K, Chandrasekhar K and Saxena, K G (2003), Indigenous ecological knowledge, biodiversity and sustainable development in the central Himalaya: *Tropical Ecology*, 44: 93-111.

Rao K S, S Nautiyal, R K Maikhuri and K G Saxena (2002), Local peoples knowledge, aptitude and perceptions of planning and management issues in NDBR, India, *Environmental Management*, **31 (2):** 168-181.

Rosenzweig C, A. Iglesias, X.B. Yang, P.R. Epstein and E. Chivian, (2001), Climate change and extreme weather events: implications for food production, plant diseases and pests. *Global Change and Human Health*, **2**, 90-104.

Saxena K G, Rao K S (2009), Climate change, biodiversity and livelihood in Indian Himalaya: The elements of uncertainty and diverse perceptions. In proceeding of Climate change data requirement and availability, Sunil Nautiyal and B. P. Nayak (eds.), Centre for ecological economics and natural resources, Institute for social and economic change, Banglore pp 78-88.

Semwal R L, Maikhuri R K, Rao K S, Singh K and Saxena K G (2002), Crop productivity under different lopped canopies of multipurpose trees in central Himalaya, India. *Agroforestry System*, 56: 57-63.

Sen K K, Rao K S and K G Saxena (1997), Soil erosion due to settled upland farming in the Himalaya: a case study in Pramati Watershed. *International Journal of Sustainable Development and World Ecology*, 4: 65-74.

Sen K K, Semwal R L, Rana U, Maikhuri R K, Rao K S and Saxena K G (2002), Patterns and implications of land use/land cover change: A case study in Pranmati watershed (Garhwal Himalaya, India). Mountain Research and Development 22, 56-62.

Sexana K G, Ramakrishnan P S, Maikhuri R K, Rao K S and Patnaik S (2004), Assessment of Vulnerability of Forests, Meadows and Mountain Ecosystems Due to Climate Change. Com-4E:\ Winrock\ Agriculture, Forestry Ch-14 pm6 2nd Proof 20-01-04. *Agriculture Forestry and Natural Ecosystem.*pp.163-168.

Singh G S, Rao K S, and Saxena K G (1997), Energy and economic efficiency of the mountain farming system: a case study in the north-western Himalaya. *Journal of Sustainable Agriculture*, 9: 25-49.

Singh K, Maikhuri R K, Rao, K S and K G Saxena (2008), Characterizing land- use diversity in village landscapes for sustainable mountain development: a case study from Indian Himalaya. *Environmentalist,* 28, 429-45.

Wallin J R and P E Waggoner (1950), The influence of climate on the development and spread of *Phytophthora infestans* in artificially inoculated potato plots. Plant Dis. Reptr. Suppl. 190. pp 19-33.

Climate change and social media: Facebook as an effective climate messenger

Rashi Mishra

School of Communication, Doon University, Kedarpur,
Dehradun - 248 001, Uttarakhand (India)
[E-mail: rashidoon@gmail.com]

Introduction

The slow process of climate change has grabbed attention as it has now become a believable fact that this change will result into massive global problems in near future, and that it constitutes a challenge for humanity to mitigate or adapt to. But even if climate change has been discussed in international meetings and conventions, various studies and surveys says that most of the people are neither very clear about climate change phenomenon nor its impact on their daily lives. Hardly people are aware of this slow change which has been occurring in the climate since long.

Scientists' and the general public often do not speak the same language when it comes to climate change and other environmental problems. Indeed, a recent public opinion poll by Villanova University shows that most people are concerned about environmental issues such as climate change, but lack a clear understanding of how these problems arise and what to do about them (Chalecki, 2000).

Climate change need to be understood beyond this as it has an impact on day to day life and is directly affecting public health. It may completely ruin human lives as it has increased health problems like asthma and allergies, speeded infectious diseases like malaria, coughing, winter weather anomalies, drought, food insecurity. List is unending but the unfortunate part is that there is no direct visibility of this climate change; people are hanging at the periphery of this hot-ball phenomenon. A cruel reality is far behind their imaginations and thoughts.

Mosquito-borne diseases are appearing higher in the mountains of Africa, Asia and Latin America, where glaciers are retreating. And the number of asthma sufferers in the U.S. has doubled since 1980 and exacerbating factors stem from burning fossil fuel (Epstein, 2011).

There is an urgent need to aware people about climate change. There should be unique methods for the same. The present paper tries to focus on new media, social media particularly social networking site 'Facebook' as an effective climate change awareness tool.

Materials and methods

Degree of media effects vary from person to person as per one's age, sex, nature, stage of life, and profession. Media influences many facets of our daily lives. Media affects our thoughts, feelings and emotions. Effect of particular media on individual depends on the use of media content by the individuals.

There are people who know and there are people who do not know about this change. But the difference lies in the ratio. Higher are the numbers who have no serious idea about climate change except "some change in the climate that is harmful" but what, when, where, why, who and how are some missing notes from an average human psyche. There comes the new media: social media.

Who knows about Climate Change; Scientists and environmental experts; and how well lay people understand the basic vocabulary of climate change. So, who should also know about climate change: Lay people (knowingly but in most of the cases, unknowingly adding fuel to the global warming). It is a well established fact that for communicating climate science, the Internet is the key.

Real-time Internet and news media communication seems to be the key. All the traditional science communication channels are proving to be slow and too easily outmaneuvered by pseudo- and anti-science narratives in the modern information environment. Unfortunately, scientists need to more actively follow up on how their own work or field is getting reported, and publicly respond to misinterpretations. Scientists need to learn the new media (Hamilton, 2012).

It is reported that knowledge of climate change among laypersons in the US, as well as in Europe and South America, is vague and contains misunderstandings (Sundblad Eva-Lotta, 2008). Still, people may not always realize this vagueness. For example, in the year 2002 the European public stated that they thought they were well informed about climate change. Ideally, people should be knowledgeable of climate change.

Another study by Ann Bostrom, M Granger Morgan: Baruch Fschhoff and Daniel read 2, people tended to confuse stratospheric ozone depletion with the greenhouse effect and weather with climate. Automobile use, heat and emissions from industrial processes, aerosol spray cans, and pollution in general were frequently perceived as primary causes of global warming. Additionally, the "greenhouse effect" was often interpreted literally as the cause of a hot and steamy climate. The effects attributed to climate change often included increased skin cancer and changed agricultural yields. Local weather effects and global warming is the result of holes in the ozone layer and attributes later to sprays such as Right Gurad which has been interpreted as aerosol spray.

Why experts should communicate about climate change to lay people. Scientists generate new findings; society relies on people using the existing knowledge to act in the common interest. Thus, it is important for the public, as well as for scientists, that current and new knowledge is communicated.

According to the Global Warming's Six Americas in May 2011 report by the Yale Project on Climate Change Communication only 39% of people in the United States believe that "most scientists think global warming is occurring", and 40% believe that "there is a lot of disagreement among scientists about whether or not global warming is happening". It is quite clear that the public opinion perception on the climate change scientific consensus is far from the true. But climate change is gaining increasing attention by decision makers and involves policy resolutions at both local and international scale. This is a main topic involving the public opinion at large as much as it is discussed in the academia. Are scientists able to communicate their knowledge to a wider audience and to avoid misunderstandings and confusion about what science really knows on the changing climate?

Traditional media like newspaper, TV, Radio, has been considered as the utmost and powerful tool for updating and informing people. Whenever any natural calamity strikes, the media gets breaking news and the country receives a huge amount of foreign aid for natural disaster relief and rehabilitation. Citizens and donors respond with swiftness and generosity. But once the tragedy ceases to dominate newspapers space and television channel slots, public interest drops steeply. Media can play a very significant role in providing complete and correct information to the people about climate changes. So far most of the science coverage has been done in the form news and features stories.

The points which are obvious from the way traditional media covers science stories: People expect the media to be sympathetic partners and allies, but the media maintain a distance, acting as observer and critics instead.

Unlike others, social media can guide people about climate change in more personal way. For population information provided by social media can be crucial in motivating and enabling them to understand climate change. Social media can stimulate a public debate in the affected communities on how to understand climate change. In her doctoral dissertation, Eva-Lotta Sundblad, 2008 "People's knowledge about climate change: Uncertainty as a guide to future commitments" Unfortunately, journalists have to deal with circumstances that can be in conflict with ambitions to provide complete scientific knowledge. For example, a journalistic text has limited space available and is normally focused on news. Consequently, there are difficulties in presenting complicated issues, which opens up for the receivers of information to develop beliefs that are distorted or simplified.

As it is difficult to define need and objective to use social media, the first question comes in the mind that how could social media act as climate messenger. People are not clear about as it has been considered by them that it fulfills only communication and entertainment part of ICE model. But it provides information, seeks feedback, medium of interaction, etc. True given the overabundance of social networking sites, it might be difficult to choose the right one.

Few questions and problems that the experts and scientists come across are: how to create and manage an account, what should be response time, what are the legal implications etc. In order to help them, Guidelines for use of Social Media have been drafted.

A group of Central Michigan University students is using social media to gather information on climate change and periodic natural events, in the Great Lakes region. Tom Rohrer, the director of the Great Lakes Institute for Sustainable Systems at Central Michigan University, and his students created a Facebook page called "Climate Change in the Great Lakes Basin". On the page students post studies, articles, pictures and other observation, which address changing weather patterns. The page is also open for the public to post their observations, creating a free and vast collection of climate change information. One among them believes that Facebook would be the best way to reach people as it is accessible to everyone and an instant communication. Using social media is also valuable because it enables the community to interact.

Discussion

If nothing else social media is used primarily to get connected and being the most popular site in India, Facebook connects! And as mentioned by Angelica Valeria Ospina in the study "Using social media to foster climate change resilience", this sense of 'belonging' and 'connectedness' to a social group can play a key role in the ability of vulnerable communities to cope with and recover from the impacts of climate change.

One cannot deny that Facebook has brought world closer and acknowledging the linkages between cultural identity and climate change resilience is particularly relevant within vulnerable developing contexts given the richness of their traditional knowledge and cultural heritage, the need for innovative responses to the challenges posed by climate change, as well as the new opportunities provided by information and communities technologies (ICTs) to access, assess and use information and knowledge. The notion of cultural identity is linked to the way in which we relate to the customs, practices, languages and worldviews that define a group or territory. It involves the conservation of social memory, the generational transfer of indigenous knowledge, the ability of a community to self organise around common interests and shared values, and the maintenance of social networks that are based on trust and solidarity, among others. All of these factors are pivotal in the capacity of vulnerable communities to deal with change and uncertainty, and to build resilience in the face of climate change.

Indians spend more time on social media particularly social networking sites than on any other activity on the Internet (http://blog.nielsen.com/nielsenwire/global/connecting-and-engaging-with-digital-indian-consumers). According to a survey conducted by networking giant Cisco (2011) in India, 95 per cent of the college students and young employees surveyed said the internet is as important in their lives as water, food, air and shelter. On the importance of social media networking sites to the younger generation, the report said 27 per cent (one in four) said they would rather update their status on Facebook than party, date or hang out with friends. In terms of usage of the social media networking site Facebook, Indian respondents were ranked highest in the survey, with 92 per cent of students checking their accounts daily, while one third (33 per cent) check their virtual walls at least five times a day. The Cisco Connected World Technology Report, 2011, conducted by third-party market research firm Insight Express surveyed 1,441 college students (aged 18-24) and 1,412 employees (aged 21-29) from 14 countries, including the US, Canada, Mexico, Brazil, UK, France, Russia, India and China.

Social media has attracted youth as they connect with friends, classmates and people with or without similar interests. Social media has its impact on the psyche of age group of 14-17 as this age group has been chosen as perceived from literature teenagers are the heavy users of social networking site and also early adopters of advanced technological applications. People within the age groups of 15-34 are the most active and engaging users. Comscore study says that the use of internet across various verticals is growing fast with social networking being the key driving factor of PC and mobile activity. Recent marketing surveys of Internet use in India point to the extensive use of social networking sites (SNS) such as Facebook among young urban people studying in schools and colleges (Cf. Datta, 2011; IAMAI). Data from Comscore estimate that Facebook's popularity among Indian online users is at a high with around 31.5 million users double of what it was in 2010. This clearly suggests that access to and use of the Internet is limited to the urban educated elite. According to data from Facebook, over 30 million are registered members of its social network; globally, Facebook claims it has over 850 million members. Other social networks like Orkut, MySpace and Linked In are way behind in terms of regular users both in India and around the world.

This period has been described as the era of social networking, collective intelligence, participation, collaborative creation, and borderless distribution. All in all, youth community is very close to social media. Since 2005, the use of SNS has risen rapidly and particularly Facebook continue to attract new users at an astounding rate. It took mobile phone 14 years to gather 150 million users, but only 5 years to Facebook (telephone: 89 years; iPod: 7 years)

The explosion of Facebook has suggested the richness of new media for effective communications (Shahkat, 2011). Based on the statistics taken from Facebook interface as at March 2009, there are 27 countries in the world with over 1 million people using Facebook. It was noted that Facebook usage in Asia is increasing quickly. In April 2006, Facebook opened its registration process to organizations, and more than 4,000 organizations joined within 2 weeks. An average of 250,000 people was registering to Facebook daily. In its factsheet posted online, Facebook describes itself as: Founded in February 2004, Facebook is a social utility that helps people.

Study by Nielsen, conducted through on an online survey on a sample population of 2000 people from all walks of life spread across top five metros and Tier 1 cities in India. The survey was administered across the country via India Speaks, a proprietary internet based panel managed by Nielsenware survey, Absolute Data Research & Analytics (2010), says that nearly 30 million Indians who are online are members of social networking sites and about two-thirds of them spend time on these social networking sites daily. More importantly, Indians spend more time on social media than they do using personal email. According to the study, an equal number spend up to an hour on social networking and email. However, while just 8 per cent spend between an hour and three hours on personal email, 20 per cent spend the same time on social media sites (Neilsen, 2009 - 2010).

According to a report released in February 2009, visits to social networking sites in India increased by 51% during 2008, to 19 million visitors in December 2008 (Comscore, 2009). During the 2009 national election, called India's "first digital election" the BJP used blogs, social networking communities, Twitter, and its Web site to reach potential supporters. Young, urban, technology-savvy first-time voters were the most targeted groups during the election campaign, speaking to the growing power of young people and the increasing importance of urban voters.

McQuail suggests that personal social circumstances and psychological dispositions together influence both general habits of media use and also beliefs and expectations about the benefits offered by the media, which shape specific acts of media choice and consumption (with consequences for further media use) and possibly, applications of benefits acquired in the other areas of experience and social activity. Psychological factors influence media behavior in two ways: they determine an individual's motivation to consume media and they also determine how an individual understands media messages and makes use of them. With the help of psychological factors, media audiences can be classified as heavy users and light users. Life style of an individual also influences media behavior because media are linked to an individual's self image as well as his social self. Life style choices decide how an individual perceives and behaves towards the media. Being a social being, man's media behavior is formed, shaped and influenced by the society. Thus interpersonal factors influence media consumption. This involves the process of socialization that is, the manner in which a society forms various guidelines for media use, like speaking about the quality of a media, social acceptability of the medium and its content.

Web 2.0 Generation survey conducted by Tata Consultancy Service (TCS, 2009) in 12 metros and mini metros across India finds that urban teenagers in the metros and mini-metros are immersed online and have the technology at hand to access information through the net at all times. Over 80% have access to mobile phones, find time for the internet alongside school classes and extra-curricular activities, and are starting to embrace Web 2.0 tools like blogs and social networking sites. Suvey also points out that Bangalore is the blogging capital of India, with 66% of students being part of blogging/social networking scene, against 39% nationally. Bangalore (9%) kids are most attracted to Secondlife, MySpace, and Podcasting, highest in India. The city has the highest (91%) iPod/ digital music player penetration nationally.... and the lowest number (15%) of students using newspaper for information access across all cities. But this survey was restricted to a privileged sector of Indian society, the urban or semi urban, English-educated young. Sarah Chong reports in Asia Tech news for the world (2009) Facebook is huge in the more developed cities of India. However, checking Emails remains to be the primary reason why Indians go on the web. Broadband disturbances and the lag in technology advancement have become stumbling blocks in the adoption of social media, but that did not stop social sites, such as Twitter from gaining ground in India.

In urban areas, Internet cafés are the primary space where first-time technology users become initiated (Rangaswamy, 2007). In his study of Bangalore Internet cafés, Nisbett (2006) found that while members of different socioeconomic classes frequented them, many used them for such mundane tasks as e-mail and Internet-related chat (IRC). Furthermore, the young men who were the immediate focus of Nisbett's study actively appropriated and shaped ICT spaces in ways that went beyond communication agendas to the acquisition of a broad range of IT skills. In rural areas, public technology access is most often provided in Internet kiosks set up by governments and NGOs; one study estimates that rural Internet kiosks could provide the first experience with ICT for as many as 700 million Indians (Rangaswamy, 2007). However, rural Internet kiosks in Tamil Nadu were mostly frequented by male high school and college students of higher socioeconomic status. This means that, unless specific steps are taken to ensure that women and other marginalized groups can visit these public spaces and take advantage of the tools provided there, the technological marginalization of these sections of the Indian population will increase further.

Social media can act as a spontaneous medium to voice sentiments and informing mass about this change in detailed way. Already several Facebook pages like *The Climate Change, Climate Change Policy & Practice, Climate Change Guide, This is Climate Change, COP 16,* are transforming the level of awareness and updating them.

This can also facilitate the sharing of thoughts and ideas between the experts and who need to know about the communities at the regional, national and global level, fostering dialogue, learning and tolerance between diverse groups.

The sense of 'belonging' and 'connectedness' that Facebook enable can contribute to reduce the anxiety and uncertainty associated with climate change impacts, as well as to improve the local responses through stronger networks, flexibility and self-organisation, among others. But how to best utilize these new forms of media to bring as much attention to the science as possible!

In the present state social media has been successful in some respects to raising awareness vis-à-vis climate change.

As a communications associate, part of this study is figuring out how to best utilize social media (Facebook) to be the climate change messenger. Emerging research and experiences suggest that Social media can play a supportive role in spreading awareness.

Conclusions

Emerging research and experiences suggest that Social media can play a supportive role in spreading awareness.

In a study done by Vani Sarraju Rao, *San Jose State University*, "Public Awareness about Global Warming in Hyderabad, India" it has come out that people showed a tendency to advocate action by society and government rather than by individual initiatives to address the problem.

In a study done by Nimish Kapoor "Role of mass media in promotion of environmental awareness along with skill development among the rural people of Shringverpur, Allahabad district, India" Nimish Kapoor discussed that different sources of mass media were not fully utilized in the studied area which hindered not only the awareness level of the respondents but also adversely affected the adoption level of new farming technologies by the rural people. There is urgent need for effective mass communication strategies to create environmental awareness and extension of agriculture based technologies among the rural people in Shringverpur village, Allahabad district in order to improve their financial status. Some functional action plans have to be chalked out and environmental awareness programmes at grass root level should be introduced in the area to educate the local-communities about the conservation of natural resources. In this condition, role of media in science communication and out-reach based agencies are very crucial to bridge the gap between science and society and efforts of these agencies may guide media industry to conceptualize and produce more television and radio programmes on success stories of eco-friendly technologies and environment related issues with their local threat and solution.

It is high time to understand that this is the time to change approach towards science; time to improve public engagement in debates on scientific issues; to develop communication strategies. It is time to move away from the traditional image of 'science and image of scientists' as a distance learning program. A need for two-way communication between the public and scientific community is needed.

Social media providers a hands-on and interactive approach that further help to break down barriers between the scientific community and the public, as well as establishing science communication as a dialogue and higher are the chances of public engagement in scientific debates.

References

Bostrom A, Morgan G M, Fischhoff B and Read Daniel, What Do People Know About Global Climate Change? 1. Mental Models *Risk Analysis,* Vol. 14, No. 6, 1994, 960-963

Chalecki E L, Same Planet, Different Worlds: The Climate Change Information Gap, Pacific Institute for Studies in Development, Environment, and Security, Influences on Public Understanding of Climate Change, 2000, A2 15-A2 16

Comscore study, 2010

Datta, Survey, Internet And Mobile Association of India (IAMAI), 2011

Epstein P, *The* Atlantic Health and Climate Change: 7 Ways You Are Being Harmed, 2011

Hamilton L, Communicating climate science, the Internet may be the key, Euro-Mediterranean Centre for Climate Change (CMCC), Climate Science and Policy, Interview was released on February 28th 2012)

Kapoor N, Role of mass media in promotion of environmental awareness along with skill development among the rural people of Shringverpur, Allahabad district, India, International Conference on Chemical, Biological and Environment Sciences (ICCEBS'2011) Bangkok Dec., 2011, 264 & 266

Nielsenwire survey, Absolute Data Research & Analytics, 2010

Neilsen, How Teens Use Media A Nielsen report on the myths and realities of teen media trends, 2009-2010

Rangaswamy, N, (2007) ICT for development and commerce: A case study of internet cafés in India, 9th International Conference on Social Implications of Computers in Developing Countries, Sao Paulo, Brazil, May 2007

Sundblad Eva-Lotta, People's knowledge about climate change: Uncertainty as a guide to future commitments Department of Psychology, Doctoral Dissertation at the University of Gothenburg, 2008, 2-4.

Shahkat S M, The Use of Facebook to Increase Climate Change Awareness among Employees, International Conference on Social Science and Humanity, IPEDR vol.5 2011.

TCS, Urban students are digital natives, reveals TCS Generation Web 2.0 survey, Tata Consultancy Services Generation Web 2.0 Survey, 2009.

Session- III

Himalya Informatics and Database

Diurnal variation in precipitable water vapour over Almora, Central Himalaya India

Sneh Joshi[1], *Kireet Kumar*[1] & *Bimal Pande*[2]

[1]G.B. Pant Institute of Himalayan Envitonment and Development Kosi-Katarmal, Almora 263 643, Uttarakhand, India
[2] Kumaun University Nainital, India
[E-mail : snehjoshi31@gmail.com]

Introduction

Precipitable water vapour (PWV) is one of the most significant elements of the atmosphere since it is the means by which moisture and latent heat are transported to cause "weather" (Gutman *et al.* 1994). Water vapour is also a greenhouse gas that plays a critical role in the global climate system. It is both a symptom and a cause of the greenhouse effect. Water vapour in the atmosphere allows the short wavelength radiation of the sun to pass through the atmosphere, but traps the long wavelength radiation emitted by the earth's surface. This trapped radiation causes the temperature to increase. As the temperature increase, the air can sustain a larger amount of water vapour, thereby increasing the greenhouse effect. Thus, it is important that water vapour content in the atmosphere is understood for the formation of climate models. Despite its importance to atmospheric processes over a wide range of spatial and temporal scales, it is one of the least understood and poorly described components of the Earth's atmosphere. Currently water vapour is measured using radiosondes and ground or space based water vapour radiometers. Radiosondes produce an accurate measurement of the water vapour profile, but the temporal and spatial resolution is rather poor (Guerova 2003). Ground based radiometers experience problems during periods of rainfall and space based radiometers can be degraded in the presence of clouds (Guerova 2003). Besides these limitations, all systems involve considerable costs. Over the past few years, a new technique to measure integrated water vapour (IWV) has been developed. This technique is based on the estimation of the tropospheric delay time of GPS signals. The delay, regarded as a nuisance parameter by geodesists, can be directly related to the amount of water vapour in the atmosphere, and hence is a product of considerable value for meteorologists. Ground based GPS water vapour estimation is not affected by rainfall and clouds. It takes minor effort to obtain GPS water vapour estimates from the existing GPS infrastructure, and since the temporal and spatial resolution is higher than of the current techniques used. Many scientists have estimated water vapour in many region of India. Jade *et al.* (2005) have estimate integrated water vapour (IWV) from GPS data of Bangalore, Kodaikanal, Hanle and Shillong over the 3-year (2001-2003). Prasad *et al.* (2007) have given that water vapour obtained from GPS data and MODIS is found to be highly variable over agriculturally very productive Indo-Gangtic plain. Kumar *et al.* (2009) examined the variability of GPS derived water vapour and MODIS water vapour over the northern Indian plains (IIT Kanpur and BHU Varanasi). Most of these studies focus on plain areas. Dai *et al.* (2002) studied the diurnal variation in atmospheric water vapor with GPS and radiosonde data over 54 stations in North America. Their results indicated that most of the stations are characterized by significant diurnal variations in precipitable water. Kimura *et al.* (1997) and Ohtani (2001) examined the diurnal cycle of rainfall and GPS-derived precipitable water around the central part of the main island of Japan. Their results showed that PWV increases in the mountainous area during daytime on calm summer days because of the convergence of anabatic winds and sea breezes.

Present study consists preliminary examination of the diurnal variation of precipitable water at one of our permanent GPS station. A distinct diurnal variation in water vapour with a maximum in the late afternoon is observed. Results show that the diurnal cycle of water vapour is related to the intensity of incident solar radiation at the surface.

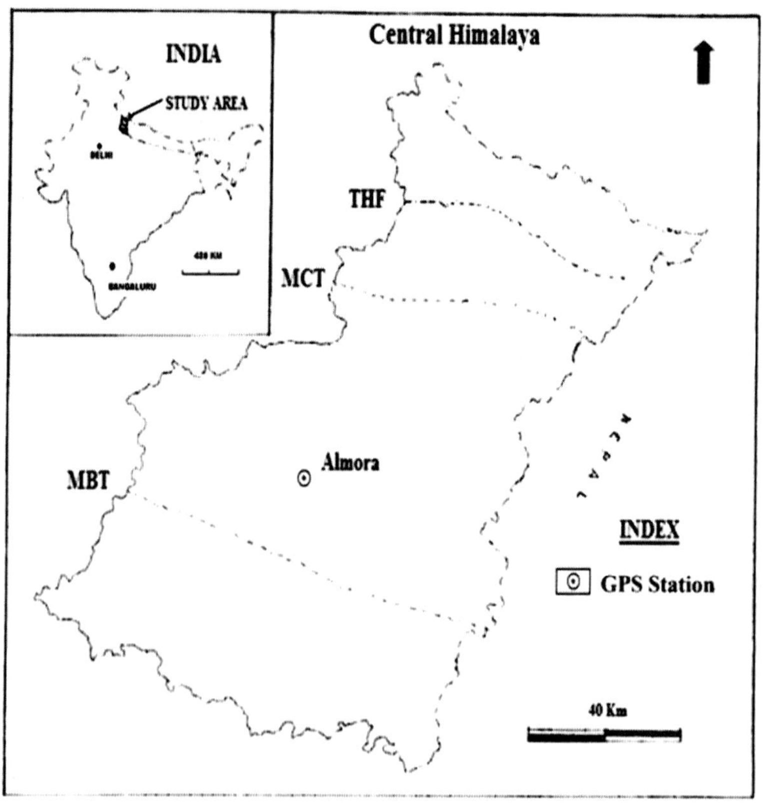

Fig 1. Study Area

Materials and methods

A permanent GPS station at Almora (29°38′ 16°72''N, 79°37' 13°08'' E and altitude 1259 m asl) has been selected for this study, which is situated in Central Himalayan Region (Fig. 1). For meteorological data collection, an automatic weather station of Cambell Scientific (Model CR10X) is installed near to permanent GPS station. The study area falls in the Lesser Himalayan domain and experience a temperate climate. Mean annual rainfall received by this region is 1100. This area is influenced by the southwest monsoon, about 70 to 80% of the total annual rainfall occurs during this period. Mean monthly temperature in summer is varies from 19°C to 24°C.

The GPS data for Almora is obtained using Trimble Survey Grade Dual frequency receiver (Trimble 5700 with chockring antenna, 1.2 and 1.5 GHz). The accuracy of GPS water vapour measurement partially depends on the accuracy of surface pressure measurements. Pressure and temperature value is obtained from Metpack sensor in rinex format with GPS data.

The data obtained in the period from January 2009 to February 2010 were processed using GAMIT software by referring to seven IGS stations (HYDE, IISC, KIT3, KUNM, LHAZ, POL2 and SELE). The atmospheric delay of GPS was estimated from GPS data. The Zenith Tropospheric Delay (ZTD) was obtained at two-hour intervals. The Zenith Hydrostatic Delay (ZHD) was estimated from surface pressure, latitude and ellipsoidal height of the observation station. The two hourly air pressure data recorded from automatic weather station of Cambell Scientific (Model CR10X) is used to calculate this dry term. The Zenith Wet Delay (ZWD) was obtained by subtracting ZHD from ZTD. The ZWD was converted into atmospheric PWV by using K parameter. The weighted mean temperature (T_m) was estimated from surface temperature (Ts). Air pressure, surface temperature, total solar radiation and rainfall data were collected from Automatic weather station. The relation between these equations is given by the followings expression:

$$ZTD=ZHD+ZWD \tag{1}$$

ZHD can be calculated from the surface pressure P_s (millibar), latitude Ø and ellipsoidal height H_s using the Saastamoinen model, which is given by the following formula (Saastamoinen 1972):

$$ZHD=(2.2779\pm0.0024)*\left[\frac{P_S}{\{1-0.0026*\cos(2Ø)-0.00000028*H_S\}}\right] \tag{2}$$

Zenith Wet Delay is calculated as:

$$ZWD = ZTD - ZHD\,. \tag{3}$$

Precipitable water vapour can be calculated from Zenith Wet Delay (ZWD) and dimensionless proportional constant K

$$PWV=K*ZWD\ ... \tag{4}$$

where, $K = [10^{-6}(\frac{k_3}{T_m} + k_2)R_v\rho]^{-1}$

where, $k_3 = (3.776 \pm 0.004)\ 10^5\ K^2m\ bar^{-1}$,

$k_2 = (17\pm10)\ Kmbar^{-1}$

T_m is the weighted mean temperature of the atmosphere, ρ is the density of water, R_v is the gas constant for the water vapour.

Weighted mean temperature of the atmosphere T_m(K) is given by following linear regression (Bevis *et al.*, 1992):

$$T_m = 70.2 + 0.72T_s. \quad (5)$$

where, T_s is the surface temperature in Kelvin. T_m is mean temperature in Kelvin, P_s is pressure of the observation site (hPa), ø is latitude of the observation site in radian, H_s is ellipsoidal height of the observation site in km.

Observed surface temperature and pressures collected from the nearby automatic weather station are transferred to the GPS stations by using interpolation technique (Berg 1948; Klein Baltink *et al.* 1999; Bai and Feng 2003).

Results and discussion

Figure 2 and 3 shows the time series of 2 - hourly GPS-derived PWV and precipitation observed at Almora from 1st June to 30th September 2009 for monsoon season and 1st December, 2009 to 28th February, 2010 for winter season. It is seen that the variation of the PWV is very large, ranging from about 5 mm to 53 mm during the monsoon month and from 0.02 mm to 19.18 mm for winter month. PWV increases before rainfall and dropped down after rainfall event. It is noted from the time series of monsoon season that PWV decreased rapidly after the end of monsoon season and continuously increased before starting of monsoon season and was the lowest in June with value 5 mm. There is a difference in change of PWV values between the winter and the monsoon season, but variation of PWV is smaller in the monsoon season.

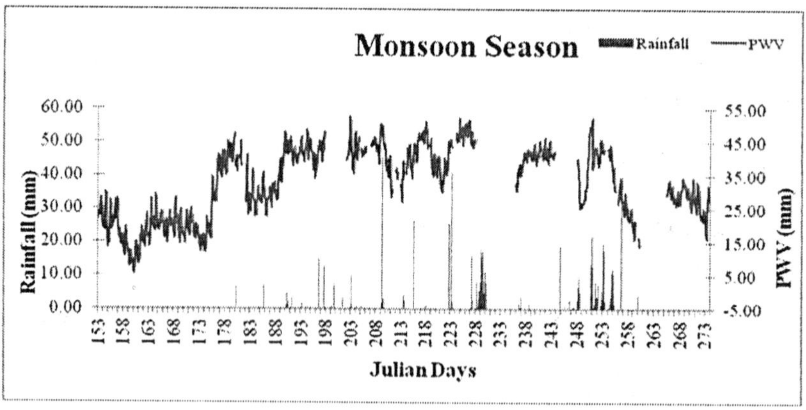

Fig. 2. Two hourly GPS derived water vapour and rainfall in monsoon season in 2009

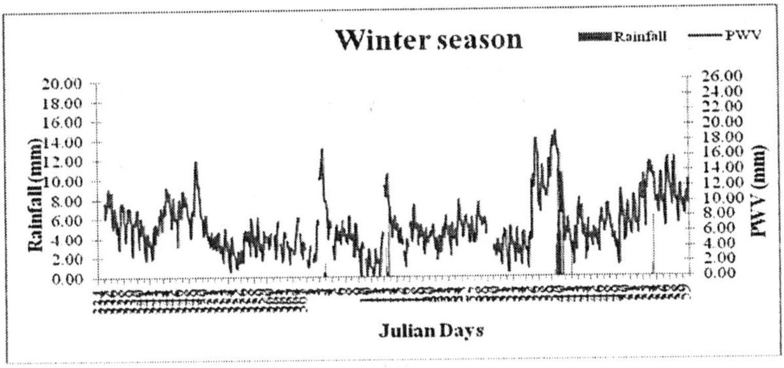

Fig. 3. Two hourly GPS derived water vapour and rainfall in winter season in 2009 and 2010

The diurnal variations of PWV in winter month (December, January and February) and in monsoon month (June, July, August and September) were examined. The PWV decreased in the morning and was the lowest at 6:00 am in monsoon months and 8:00 am in winter months. It increased in the afternoon, was maximum and high value of PWV in late afternoon in all months of both season (Fig. 4 and 5). The increase of PWV during daytime is mostly due to evaporation of water vapour from the surface. The air temperature increases from 6:00 am reaches its maximum at around 12:00 noon. But the PWV value did not show corresponding increase with the air temperature in the morning. This may be caused by PWV augmentation by the water vapour transport through the local circulations, which reduce the possible increase of PWV with temperature rise. The increase of PWV during daytime and its maximum value in the late afternoon are similar to the results of a numerical model study done by Kimura and Kuwagata (1995) by simulating the thermally induced local circulation. Diurnal variations between PWV and rainfall for Almora explain that at hourly scale, rainfall events precede PWV and the PWV values goes down after the rainfall events (Fig. 6 a, 6b, 6c and 6d). In July, August and September rainfall has clear diurnal cycles, but the PWV changes little during the day. It is indicated that PWV shows little change when it is close to its saturated value. Diurnal variations in June are different. In this month diurnal cycle of water vapour is clear because this is monsoon initiation month and cloud activity is low in this month.

Mean diurnal variations in PWV is also analyzed for low (<12.0 MJ/m^2) and high solar radiation days (>18.0 MJ/m^2) for monsoon season. Value of mean PWV is high in case of low solar radiation days but amplitude of PWV is smaller on the days of lower total solar radiation. Figure 7 and 8 show the mean diurnal variations of PWV for days categorized as high and low total solar radiation days. The lower solar radiation means sky is with thicker and spatially extensive clouds. This result shows that the amplitude of PWV is greater in the case of high total global solar radiation. The daily maximum value appears at different times in these two cases. Low solar radiation reduces the intensity and thickness of the thermally induced local circulations so that the convergence of the low-level winds around the mountains is reduced. High solar radiation increases the intensity and thickness of local circulation, especially with respect to the larger horizontal-scale circulations, which would sustain the low-level convergence until the late afternoon (Wu *et al.* 2003).

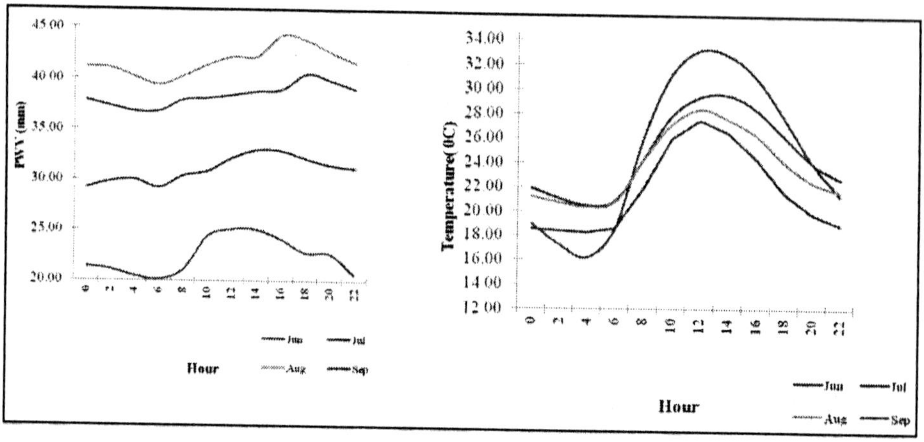

Fig. 4. Diurnal variation in PWV and Temperature in monsoon month

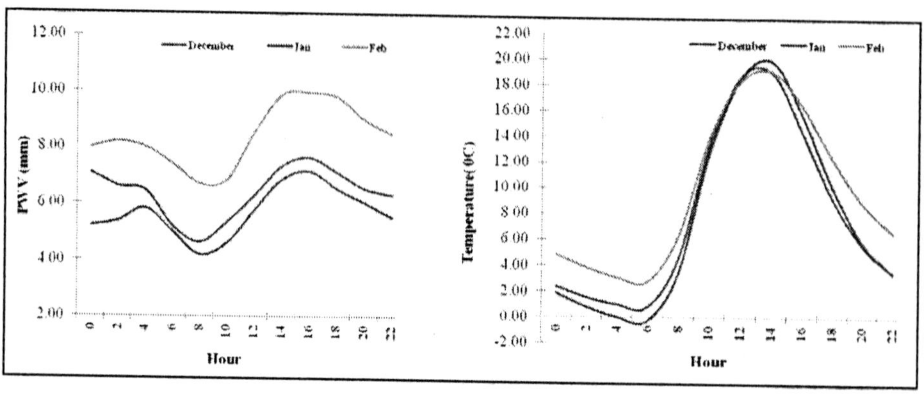

Fig. 5. Diurnal variations in PWV and Temperature in winter month

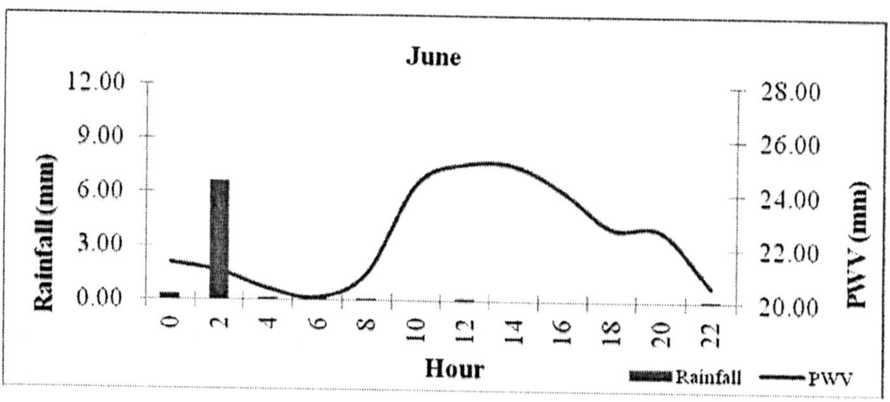

Fig. 6a. Diurnal variation in PWV and Rainfall in June

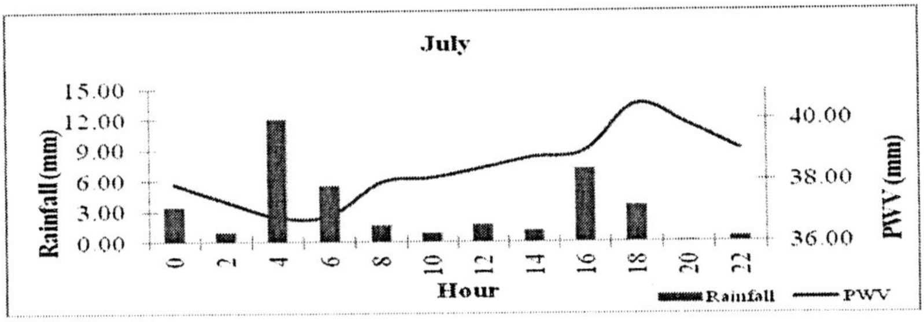

Fig. 6b. Diurnal variation in PWV and Rainfall in July

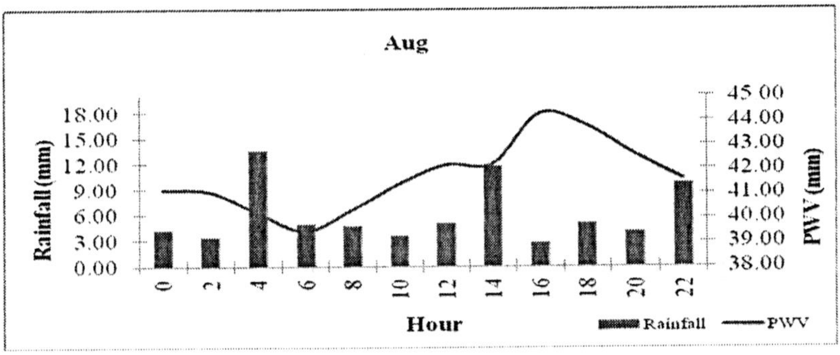

Fig. 6c. Diurnal variation in PWV and Rainfall in August

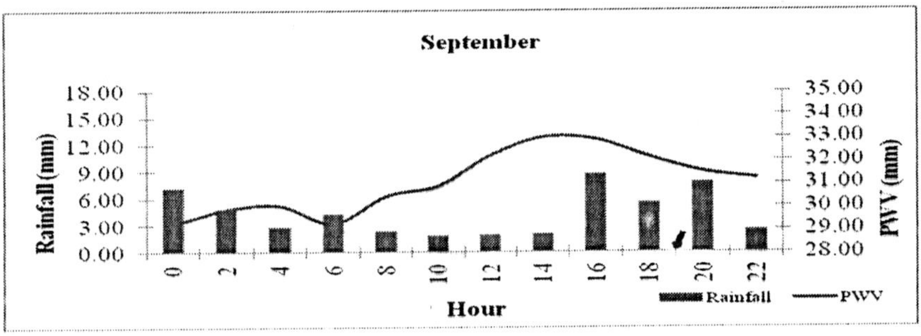

Fig. 6d. Diurnal variation in PWV and Rainfall in September

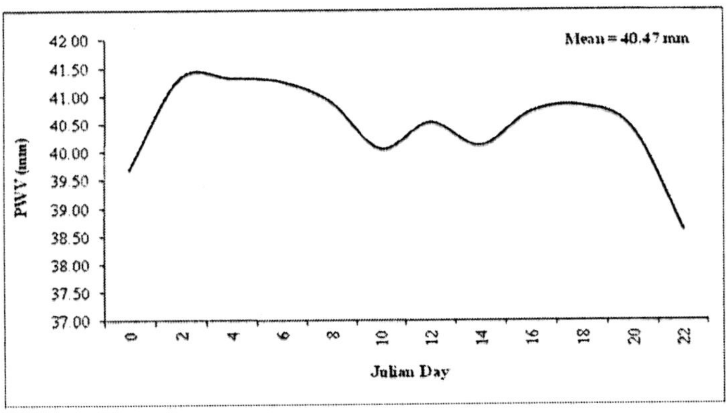

Fig. 7. Mean diurnal variations of PWV on low solar radiation days

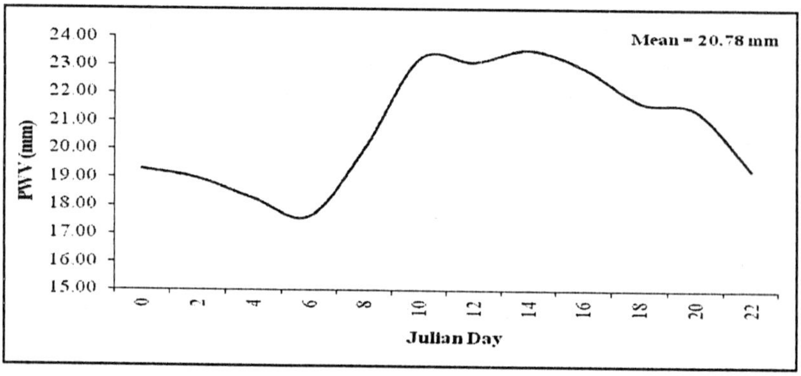

Fig. 8. Mean diurnal variations of PWV on high solar radiation days

Figure 9 and 10 shows the mean diurnal variations of PWV on days with heavy rain (Rainfall > 10 mm) and days without rain during monsoon season. The results show that the amplitude of PWV on days with heavy rain is larger than that on days without rain. The maximum value of PWV on days with heavy rain appears 2 hour later than that on days without rain. Cloud activities are very strong on days with heavy rain and thermally induced local circulations reduced on these days so very low diurnal variation are recorded in heavy rainfall days.

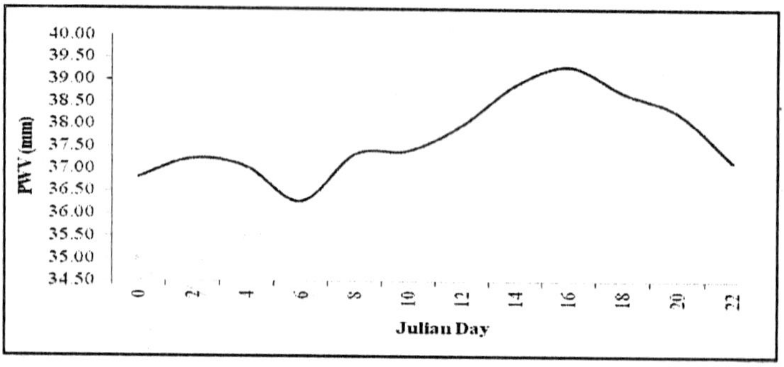

Fig. 9. Mean diurnal variations of PWV on days with heavy rain (total rainfall > 10 mm)

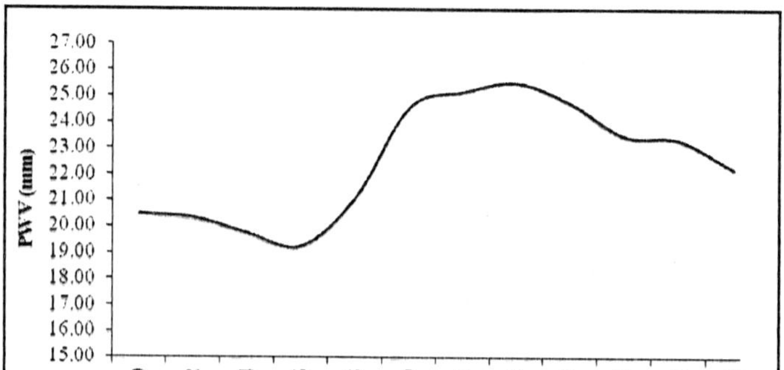

Fig. 10. Mean diurnal variations of PWV on no rainfall days

Conclusion

Knowledge of spatial and temporal variation in atmospheric water vapour over the Himalayan region is important in forecasting regional weather and for understanding the global climate system. Precipitable water vapor was estimated from GPS observation at Almora of Central Himalay India. The GPS data obtained from January 2009 to February 2010 with high time resolution. PWV value was different during winter and monsoon months. The diurnal variations of PWV in winter and monsoon month were investigated. The PWV was minimum at 6-8 am in the morning; it increased in the afternoon and kept with the high value from the evening to the midnight. This may be a result of evaporation from the earth's surface and water vapor transport by the local circulation. The diurnal range of the PWV on days with heavy rain was larger than that on days without rain. Value of mean PWV is high in case of low solar radiation days on monsoon season. This paper is preliminary study of PWV in monsoon and winter month based on one year data. To explain the annual, seasonal and diurnal behaviour of precipitable water vapour and other meteorological parameters over the Central Himalaya, study will be investigated on the long term data.

Acknowledgement

Authours are thankful to Ministry of Earth science (MOES), India for funding for the GPS data of Indian GPS networks and to the Director, G. B. Pant Institute of Himalayan Environment and Development, Kosi-Katarmal Almora for facilities and help. We would also like to acknowledge the support given by Dr. R.W. King, Department of Earth, Atmospheric, and Planetary Sciences, Massachusetts Institute of Technology.

References

Bevis M, Businger S, Herring T A, Rocken C, Anthes R A and Ware R H (1992), GPS Meteorology: Remote sensing of atmospheric water vapour using the Global Positioning System. J Geophys Res 97: 15,787-15, 801.

Berg H (1948), Allgemeine meteorology. Dümmlers, Bonn.

Bai Z and Feng Y (2003), GPS water vapor estimation using interpolated surface meteoro-logical data from Australian automatic weather stations. Journal of Global Positioning Systems 2: 83-89.

Dai A, Wang J, Ware R H and Hove T Van (2002), Diurnal variation in water vapor over North America and its implications for sampling errors in radiosonde humidity. J. Geophys. Res. 107: 4090.

Gutman S I, Chadwick R B, Wolf D, Simon W A and Hove T Van and Rocken C (1994), Toward an Operational Water Vapor Remote Sensing System Using the Global Positioning System. Proceedings of the Fourth Atmospheric Radiation Measurement (ARM) Science Team Meeting. DOECONF-940277, March 1994 Charleston, South Carolina.

Guerova G (2003), Derivation of Integrated Water Vapor (IWV) from the ground – based GPS estimates of Zenith Total Delay (ZTD). IAP Research Report No. 2003-08. Dept. of Microwave Physics, Institute of Applied Physics, University of Bern, Bern, Switzerland.

Jade S, Vijayan M S M, Gaur V K, Prabhu T P and Sahu S C (2005), Estimates of precipitable water vapor from GPS data in Indian subcontinent. J Atmos Sol Terr Phys 67(6): 623-635.

Klein Baltink H, Derks H J P, van Lammeren ACAP, Ambrosius B A C, Van der Hoeven AGA, van der Marel H, Kleijer F, and Kösters A J M (1999), GPS Water Vapor Meteorology, Beleids Commissie Remote Sensing (BCRS), Chapter 2: Water Vapor from GPS Troposheric Delay Estimates, pp.3-11.

Kimura F and Kuwagata T (1995), Horizontal heat fluxes over complex terrain computed using a simple mixed-layer model and a numerical model. J. Appl. Meteor. 34: 549–558.

Kumar S, Singh A K, Prasad A K and Singh R P (2009), Annual variability of water vapor from GPS and MODIS data over the Indo-Gangetic Plains. J Ind Geophys Union 13(1): 17-23.

Prasad A K, Singh R P, Singh S and Nanda D S (2007), GPS and satellite meteorology for understanding monsoon dynamics over the Indian subcontinent. Proceedings of Symposium HS 2004 at IUGG 2007, Perugia, July 2007, IAHS Publ. 313, pp 33- 39.

Saastamoinen J (1972), Atmospheric correction for the troposphere and stratosphere in radio ranging of satellites. In: Henriksen SW, Mancini A. Chovitz BH (eds.). The Use of Artificial Satellites for Geodesy in Geodesy, Geophys. Monogr. Ses. 15, Amer Geophys Union, pp 247-251.

Wu P, Kimura F, Sasaki T, Liu J and Yamanaka M D (2003), Diurnal variation of GPS-derived precipitable water over the Tibetan Plateau during summer.

Institutional capacities for sustenance of Himalayan ecosystem

Rajesh Kumar, Anand Kamavisdar, Nisha Mendiratta & Akhilesh Gupta*

Climate Change Programme Division, Department of Science and Technology,
Technology Bhawan, New Mehrauli Road, New Delhi 110 016, India
[E-mail: rajesh38@hotmail.com, rajeshkr38@nic.in]

Introduction

Himalayan ecosystem is rich and fragile. Perennial rivers of north India depend heavily upon the sustainability of glaciers and the ecosystem of the Himalayan region. Sustainability of an ecosystem demands a balance and equilibrium among various forms of life and their surroundings established over long periods of time. Himalayan ecosystem includes over 51 million people who practice hill agriculture and remains vulnerable[1-3]. Himalayan ecosystem is vital to the ecological security of the Indian landmass, through providing forest cover, feeding perennial rivers that are the source of drinking water, irrigation, and hydropower, conserving biodiversity, providing a rich base for high value agriculture, and spectacular landscapes for sustainable tourism. The Himalayan eco system is vulnerable and susceptible to the impacts and consequences of

- Changes on account of natural causes,
- Climate change resulting from anthropogenic emissions and developmental paradigms of the modern society.

National Mission for Sustaining the Himalaya Ecosystem (NMSHE)[4] launched by Government of India under National Action Plan for Climate Change (NAPCC) proposes to devise a mechanism to understand complex processes affecting the Himalayan Ecosystem. The mechanism evolve suitable management and policy measures for sustaining and safeguarding the Himalayan eco-system by human and institutional capacity building in different domains, networking of knowledge institutions engaged in research. This will lead to development of a coherent data base on ecosystem, detecting and decoupling natural and anthropogenic induced signals of global environmental changes in mountain ecosystems. Studies on traditional knowledge systems for community participation in adaptation, mitigation and coping mechanisms inclusive of farming and traditional health care systems and developing regional cooperation with neighboring countries enhance the mission objectives[5]. Under NMSHE it is proposed to generate a data base through monitoring and analysis, to eventually create a knowledge base for policy interventions.

Materials and methods

National mission for sustaining the Himalayan ecosystem (NMSHE)

Primary objective of the mission is to develop a sustainable national capacity to continuously assess the health status of the Himalayan Ecosystem and enable policy bodies and States to implement actions selected for sustainable development[4]. The NMSHE attempts to address the following important issues.

- Himalayan Glaciers and the associated hydrological consequences,
- Extreme weather events
- Biodiversity conservation and protection,
- Wild life conservation and protection,
- Traditional knowledge societies and their livelihood and
- Planning for sustaining of the Himalayan Ecosystem.
- Vulnerability, resilience and impact studies related to socio-economic sectors like water, agriculture, health, forest, environment, etc

Several knowledge institutions from various ministries, community based organizations and experts from the university sector are already engaged in some of the work elements associated with the mission. A strong and effective coordination among the existing knowledge institutions is being planned. These organizations will be networked and further strengthened. A Major work element under the mission should be the enrolment of partners and partnering institutions. Thematic working groups and knowledge network for each theme area will be formed from the network[6]. Mapping of eco system, monitoring of glacier resources, database on traditional knowledge systems and livelihood management practices etc. call for a partnership between governmental and community based as well as non-governmental organizations[7]. The organizational structure[4] for the NMSHE to work on the thematic area is given in Fig. 1.

Discussion

Research initiative by institution on Himalayan ecosystem

NMSHE provides an integrating platform for several knowledge institutions[8] engaged in the study of glaciers, rivers, forest, wildlife and of the other ecosystem components. The strategic knowledge generated through research work network will strive to provide inputs for evolving policy strategies and recommending action programmes for sustaining the eco system[9]. The research network will strengthen the ongoing activities in the knowledge domain by enrolling knowledge institutions in the area and develop new institutional mechanisms needed for a long term coordinated action programme through appropriate steering and monitoring mechanisms[10-12]. NMSHE designs an integrated approach for identifying vulnerabilities of the resource system to climate and socio-economic changes, and for evaluating adaptation options[5]. The major research initiative[8] with existing institutional mechanism under the mission are summarised; Inter-University Consortium on Cryosphere and Climate Change (IUCCCC) is being formulated to look into Cryosphere-Societal interactions, within the framework of integrated science-social science research. Initially IUCCCC has been constituted with four Universities, namely; Jawaharlal Nehru University, New Delhi, University of Kashmir, Srinagar, University of Jammu, Jammu and Sikkim Central University, Gangtok, having subject experts and infrastructure facilities. The consortium proposed to undertake research on glaciological studies and requires the analysis and interpretation of the myriad data on various aspects of the Himalayan cryosphere dynamics. The data thus generated would help in formulating action plan on climate and climate change impact which needs necessary mitigation measures for the future. These historical data will be useful for similar cryosphere studies[13] for other regions as well.

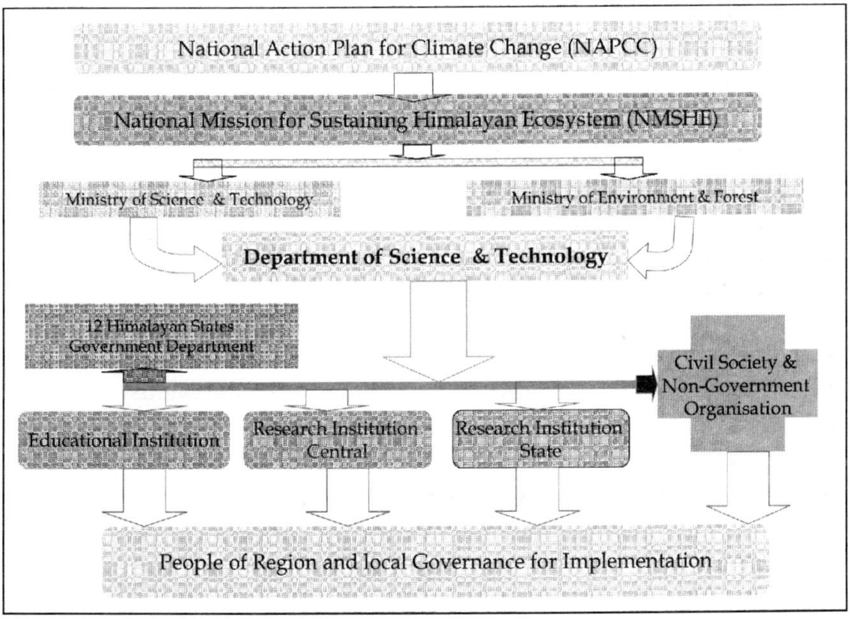

Fig. 1. Organisation structure on sustaining Himalayan ecosystem under NMSHE

Integrated Research and Action for Development (IRADe) explore the work on vulnerability due to Climate Change through scientific assessment in state of Himachal Pradesh. Some districts are vulnerable in terms of water resources, agriculture and some areas are prone to hazards such as floods, excessive rainfall, landslides and so on.

The bio-diversity study on Climate change, Aerobiology and Public Health in North-eastern Himalayan region of India is proposed by North Eastern Regional Institute of Science and Technology. The study relates to human health in the projected scenarios of climate change and needs to do analysis with regard to the projected climate change when such baseline information from earlier climate regimes are not available for the area. The North-Eastern Himalayan region identify the need of opportunities for ascertaining variation in the immune response of various tribal populations existing in the region.

Assam Agriculture University (AAU) identified the need for collection and documentation of Assessment of rice biodiversity to ascertain gaps in the existing rice germplasm collections. The work targets to fulfill such lacuna to strengthen the existing germplasm collections of wild rice varieties that may be of potential value in developing new rice varieties for overcoming adverse impact of climate change on the yield of rice. Assam Agriculture University would also work to address the traditional knowledge about selected minor fruits of the region.

A study on range shifting of distributional pattern of aquatic diversity in varied altitudes of a lotic ecosystem in Assam as proposed by Institute of Advanced Study in Science & Technology (IASST) has great potential for climate change research. The studies on the food chain relationships will be of help in interpreting the data on taxa due to indirect effects of climate change on the specific indicator species.

North East Institute of Science and Technology (NEIST) proposed study on conservation of medicinal, nutritional, aromatic and spice plants associated with the livelihood for sustainable use and to enhance socio-economy of the people in Arunachal Himalaya. NEIST is also working on characterization of the degraded lands, rattan cane productivity sustenance through appropriate inputs and agro-management, climate variability and interaction with other biotic/ abiotic stresses effects. The research work will deliver development of transfer functions and technical coefficients for response evaluation, linking CC scenarios for vulnerability, mitigation and adaptation strategies. Nagaland Science & Technology Council works on identification and documentation of Indigenous knowledge on agricultural practices in Nagaland and also biodiversity mapping and conservation at copper hills, Phek district. State Science Council, Mizoram identified need for documentation of traditional knowledge regarding the use of various herbs and plants used as medicines in Champhai district of Mizoram and to assess threatened Fauna and development of practical approach for their conservation in Bualpui V Mizoram.

Dr Y S Parmar University of Horticulture & Forestry is working on Briquetting of pine needles for power generation and environment upgradation and its regional impact on environment & GHG control. The university also proposes for documentation and conservation of medicinal and aromatic plants under changing climate scenarios mid hill zone of HP. Institute of Himalayan Bioresource Technology (IHBT) research work on assessment and prediction of changes occurrence in western Himalayas eco-system including tree line ecology and medicinal plants observe importance in the climate change scenarios. The research outcome will provide suitable mitigation and adaptation strategies, developing forecast for suitability of land use types and planning. A study on impact of climate change on the dynamics of mountain ecosystem, natural

57.		Carnivores and their prey in the context of impacts of climate change/variability on key wild animal species and their habitats in the western Himalaya.
58.		CC Impact and hydropower developments on Himalayan river basins.
59.	North Eastern Regional Institute of Science and Technology, Arunachal Pradesh	Traditional knowledge of wood crafts and carving among different tribes of Nagaland.
60.	BN College of Agriculture Biswanath Chariali, Assam-	Extreme weather events and climate change on agriculture and studies on climate resilient bio-resources/technologies for rice-based cropping system in the foot hills of Eastern Himalayan.
61.	CSIR-North East Institute of Science & Technology, Jorhat Assam	Impact assessment of functional ecosystem and services in trans-border areas of Himalayan foot hills region of Assam–Arunachal Pradesh and Bhutan
62.	Assam Agricultural University Jorhat	Carbon sequestration and crop adaptation strategies for climate resilient agriculture in unutilised and degraded land of sub-Himalayan ecosystem of north eastern region of India through rattan cane plantation
63.	Sher-e-Kashmir University of Agri-cultural Sciences & Technology of Jammu (SKUAST-J), J&K	Impact of biochar amendments on carbon sequestration and erosion mitigation under different land-use systems in foothills of Shivaliks in North Western Himalayas
64.	Wildlife Institute of India, Chandrabani, Dehra Dun.	Species susceptible to climate change impact with a view to reduce extinction risk.
65.	G. B. Pant Institute of Himalayan Environment & Development, Almora	Forest phenology based indicators on climate change using satellite remote sensing for ecosystem & regional modelling
66.	Science & Technology Council, Arunachal	Traditional Knowledge regarding the use of various Herbs and Plants used as medicines in Champhai
67.	Indian Institute of Technology, Khargpur	Ecosystem Ecology in Eastern and Western Himalayan Elevation Gradients
68.	TERI-Northeastern Regional Centre Guwahati, Assam	Bamboo Plantations in Riverine areas for livelihood generation and ecosystem development in selected districts of Assam.
69.	G.B. Pant Institute of Himalayan Environment and Development,Garhwal	Assessment of climate change impact on agro- and forest ecosystems and community response, and adaptation in Central Himalaya
70.		Climate change impact on agro- and forest ecosystems and community response, and adaptation in Central Himalaya
71.	Chaudhary Sarwan Kumar Himachal Pradesh Krishi Vishvavidyalaya, Palampur	Emerging pests and diseases of important crops and animals for Himalayan Ecosystem vis-à-vis climate change Augmenting livestock production of Himalayan Ecosystem vis-à-vis climate change

Creation of new knowledge Institute

NMSHE will activate the knowledge institutions in the region in the consolidation of fragmented efforts and mount strong and long range programmes in the areas such as[4];

- Forest wealth, biodiversity and flora/fauna,

- Weather observational network,
- Ecology and eco tourism services,
- Traditional knowledge systems.

The mechanism will focus on creation and strengthen institutions in gaps areas for virtual knowledge networks and existing knowledge institutions. The National Center for Himalayan Glaciology (NCHG) has been created by Department of Science & Technology to undertake both intra mural research and support extra mural research from other knowledge institutions. Similarly, institutional capacities in the areas of traditional knowledge systems, Himalayan agriculture, eco tourism and biodiversity would also be attempted.

NMSHE propose to establish a live portal to interconnect all states in the Himalayan Region. The portal will boost linkages between institution-community based organisations for public outreach of climate change and ecosystem related informed knowledge. The online interactive platform will increase the community participation of people living at high altitude and defence personnel.

International cooperation

Bi-lateral cooperation with Switzerland, Nepal and some other neighboring countries are proposed to emerge under the mission for building up capacity for scientific manpower. The Indo-Swiss Joint Working Group on Glaciology has taken up this critical need and recommended the initiation of Capacity Building Programme on Glaciology, comprising of three levels. The Indo-Swiss Joint Working Group (JWG) on Glaciology which met in Switzerland in May 2012 recommended for develop and organise host capacity building programme on Glaciology and related areas, under the aegis of Department of Science and Technology, Government of India. The resource persons from India and Switzerland will visit to train Indian students and young researchers in this subject area. The first module of training programme as Level I has been designed wherein 20-30 students/researchers will be trained in a four week on long basic theory and practical sessions in various fields of research related to glaciology.

Conclusions

NMSHE provides a research platform for more than 100 institutions and 500 scientists, who are working to strengthen institutional capacities for sustenance of Himalayan ecology that is proposed to be compiled. The research work from few such institutions would be initiated and compiled in coordinated manner for the sustenance of Himalayan ecosystem. A comprehensive and coordinated research work like one proposed under NMSHE network will boost the outcome and enrich research work. The mission in coordination with state Government will interconnect research efforts among enrolled knowledge institutions, community based organisations, stake holders and international agencies.

References

Anonymous, Mission Document on National Mission for Sustaining the Himalayan Ecosystem (NMSHE); Published by Department of Science & Technology, June, 2010

Anonymous, National Action Plan on Climate Change. Prime Ministers Council on Climate Change, GOI, New Delhi. 2008.

Anonymous, National Environment Policy, Ministry of Environment & Forests, GOI, New Delhi. 2006.

Anonymous, Report of the " Stake holder Workshop on National Mission for Sustaining Himalayan Ecosystem" during 6-7th October, 2012

Anonymous, Report of the Taskforce on Mountain Ecosystems – Environment & Forest Sector for Eleventh Five Year Plan, Planning Commission, GOI, New Delhi. 2006.

Gowhar Meraj, Shakil A Romshoo and A R Yousuf (2012), Geoinformatics Approach to Qualitative Forest Density Loss Estimation.-A case study of Pir Panjal Range. International Journal of Current Research and Review, Vol. 4, No. 16: 47-61. ISSN No. 0975-5241

Kamavisdar Anand, Kumar Rajesh, Mendiratta Nisha and Gupta Akhilesh, "National Initiatives to Build Strategic Knowledge on Climate Change" during National Conference on "Sustainable Agriculture & Food Security: Challenges in Changing Climate" during 27-28 March, 2012

Kumar Rajesh, Kamavisdar Anand, Mendiratta Nisha and Gupta Akhilesh, "National Initiatives for Sustenance of Himalayan Ecology" during National Conference on "Sustainable Agriculture & Food Security: Challenges in Changing Climate" during 27-28 March, 2012

Mehnaz Lone, Mehjoor Lone and Shakil A Romshoo (2011), Geospatial tools for assessing land degradation in Budgam district, Kashmir Himalaya, India. Journal of Earth System Science, Vol.120(3):423-433

Mendiratta Nisha, Kumar Rajesh, Kamavisdar Anand, and Gupta Akhilesh, "Climate Change and India's Sustainable Development Pathway" during National Conference on "Sustainable Agriculture & Food Security: Challenges in Changing Climate" during 27-28 March, 2012

Ganjoo R K and S B Ota (2012), Mountain environment and early human adaptation in NW Himalaya, India - A case study of Siwalik Hill Range and Leh valley. Quaternary International 269, 31-37

Raman R and Punia M (2012), The application of GIS-based bivariate statistical methods for landslide hazards assessment in the upper Tons river valley, Western Himalaya, India. Georisk: Assessment and Management of Risk for Engineered Systems and Geohazards, Vol. 6 (3):145-16.

Sharma M, Deswal S and Kumar P (2009), Himalayan Glacier and Climate Change: A Case of Misunderstood Science. Think India Quarterly, Vol 12/4, pp 80-87.

Effect of MSW leachate on the Deepar beel ecosystem: A ramsar site in Assam, India

Sonali Borpatra Gohain & Sabitry Bordoloi

Life Sciences Division, Institute of Advanced Study in Science and Technology, Guwahati - 781 035 Assam, India
[E-mail: sonubg@gmail.com]

Introduction

The problem of improper solid waste management has been a major environmental challenge in many Indian cities. The leachate percolating from a MSW disposal site is the major source of contamination of the surrounding ecosystems. Physical, chemical and microbial processes transfer pollutants from the waste material to the infiltrating water, resulting in a contaminated liquid containing high concentrations of organic and inorganic contaminants (Jorstad L B, Jankowski J *et al.* 2004).

The six year old (6th July, 2006 - 2012) open and unsorted Municipal solid waste (MSW) dumping site of Guwahati city (26°06′49.3″N 91°40′56.1″E) is in the fringe area of Deepar Beel, a Ramsar site (19.08.2002). Deepar Beel is home to many migratory birds and aquatic biodiversity. The Beel was currently estimated to be 9.27 sq km. However, the actual water body has now reduced to 4.1 sq km. Depth of the Beel ranges from about 6 m to 1.5 m depending on the season , monsoon or dry season. The site reportedly supports a number of IUCN red-listed species (Planning commission, *et al.* 2008). The Deepar beel provides livelihood to its neighbouring villages primarily in terms of fishing activities and water use for various purposes. The Deepar Beel is reported to provide, directly or indirectly, its natural resources for the livelihood of fourteen indigenous villages (1,200 families) located in its precincts (Roy, Kalita, *et al.* 2011).

Open dumping causes some long term residual effect on the ecology of the water bodies present in the vicinity of the site. The effect of uncontrolled infiltration of leachate into the environment is the biggest environmental impact that the open dumping site has. Discharge of potentially toxic trace elements from the leachate into aquatic ecosystems has become a serious threat. All heavy metals are potentially harmful to most organisms at some level of exposure and absorption. Under certain environmental conditions, heavy metals may accumulate to a toxic concentration and cause ecological damage (Ramos, Fernandez *et al.* 1999). Occurrence of heavy metals in the aquatic environment is of major ecological and public health concern because of their toxicity at certain concentrations, translocations through food chain, non biodegradability, which is responsible for their accumulation in the biosphere, threat to human life and the environment (Eneji, Sha'Ato, *et al.* 2012). Hence, trace metals may be sensitive indicators for monitoring changes in the water environment (Mohiuddin, Otomo, Ogawa, *et al.* 2012).

Materials and methods

Surface water samples were collected on a monthly basis, during the period from March 2011 to February 2012 from near (<20m) the Municipal solid waste dumping site. An uncontaminated site (at a distance of 5 km) was taken as the control. Water samples for trace element analysis were collected in acid-washed 250ml plastic bottles. Samples were acidified by the addition of 5ml of conc. nitric acid to maintain pH<2 and stored at 4°C in a refrigerator until analysis. 100ml sample was taken from the acid preserved sample. 5 ml conc HNO_3 was added to it. The samples were then made to evaporate in a hot plate to the lowest volume possible (about 30 ml) before precipitation occurs. A light coloured completely clear solution is obtained at the end of the digestion process. After cooling of the digested samples, the samples were then filtered using Whatman filter paper (25 mm). The filtrate is then transferred to a 100 ml volumetric flask. It is diluted to mark. The digested samples were then analysed for Zn, Mn, Cu and Pb in Atomic Absorption Spectrophotometer, iCE 3000 series (APHA, APHA *et al.* 1998). Stock standard solution of Zn, Mn, Cu and Pb were obtained from Merck in concentration of 1000mg/l. Calibration standards were run regularly to ascertain the accuracy of the analytical procedure. Mili Q water was used for all purposes.

Fish samples of *Anabus testudineus* have been collected from around the dumping site seasonally for a period of 1 year. Control sample was collected from an uncontaminated water body situated at a distance of 25 km. For analysis, the muscle tissue of the fish samples was processed and as had for assessing trace elements with AAS. About 0.1g of ash was dissolved in 5 ml of conc. HNO_3 and made up to 20ml with distilled water. The mixture was heated on a plate until obtaining the 3rd of total volume. The solution was then filtered and trace elements Zn, Mn, Cu and Pb were determined by Atomic Absorption Spectrophotometer, iCE 3000 series.

Results and discussion

The results of the physic-chemical parameters of the sampling site have been summarized in Table 1. The surface water temperature fluctuates between 21-29.3°C with a mean value of 26°C. The pH ranged from 6.9-7.4 with a mean value of 6.91. Water was found to be slightly acidic in nature. The free CO_2 range was within (10.3-14.1 mg/l) with mean value of 11.120 mg/l. High concentration of free CO_2 throughout the monitoring period might be due to the heavy inflow of organic waste in the leachate (Kosygin, Dharmendra *et al.* 2008). Dissolved oxygen has been found to be low throughout the study period ranging from (0.9-2.4 mg/l) with the mean value of DO at 1.4mg/l. The range of hardness of water has been found to be (120-324 mg/l) with mean value of 198mg/l. Electrical conductivity has been high ranging from (405-1601.1 μS). The mean value of EC recorded during the study period was 623.21 μS. The range of alkalinity in the water samples was found between (75-498 mg/l) with mean value of 160 mg/l. Total dissolved solids ranged from (105.9-446.3 ppm) with mean value at 345.3. Low DO value (0.9-2.4 mg/l) accounts for survival of organisms adapted to the conditions or organisms having accessory respiratory organs viz. air breathing fishes.

Table 1. Physico-chemical characteristics of a point nearest to the dumping area in Deepor Beel

Parameter	Range (Mean)
Water Temp. (°C)	21-29.3 (26)
pH	6.9-7.4 (6.91)
Free CO_2	10.3-14.1 (11.12)
Dissolved oxygen	0.9-2.4 (1.4)
Hardness	120-324 (198)
EC(μS)	405-1601.1 (623.21)
Alkalinity	75-498 (160)
TDS	105.9-446.3 (345.3)

Table 2. Concentration of trace elements in water and fish in Deepor Beel

Trace element	Water (contaminated site)	Water (control)	Fish (Experimental)	Fish (Control)
Manganese				
Range	0.514-1.5314	0.0712-0.2536	2.659-4.844	0.1182-2.010
Mean	(1.2314)	(0.43549)	(3.809)	(1.9859)
Zinc				
Range	0.171-0.947	0.1009-0.3960	2.100-3.954	0.4820-0.9118
Mean	(0.5902)	(0.18705)	(3.712)	(0.7625)
Copper				
Range	0.008-0.312	0.0012-0.0459	0.048-0.412	BDL
Mean	(0.0815)	(0.04309)	(0.211)	
Lead				
Range	0.010-0.118	BDL	0.19-0.238	BDL
Mean	(0.0110)		(0.204)	

The range of the concentration of trace elements in the water samples and in fish tissue along with control samples during the entire study period is given in Table 2. The trace element in water samples were found to occur in the decreasing sequence of Mn >Zn>Cu>Pb. The concentration of Mn in the water samples was found in the range of (0.5149-1.5314 ppm) with the mean concentration at 1.2314 ppm. The Zn concentration in water ranged from

(0.1712-0.9472 ppm) with mean concentration of 0.5902 ppm. The Cu concentration ranged from (0.008-0.312 ppm) with mean concentration of 0.0815 ppm. For Pb, the concentration range was between (0.01-0.118 ppm) with mean concentration of 0.0110 ppm. The mean concentrations for Mn, Zn and Cu were found within their permissible limits for surface water as given by CPCB (1986) (Kosygin L, Dharmendra H, *et al.* 2008). But the mean concentrations of the trace elements Mn, Zn and Pb exceeded their limit for drinking water as given by WHO (1993) (Table 3). The Deepor beel supports a number of neighbouring villages where people use the beel water for their daily activities like drinking, cooking and washing clothes and utensils. The beel water is found to be unfit for such human activities as per WHO guidelines for drinking water (1993) (Jarup L, 2003). Therefore, such usage might cause harmful implications on the human habitation in the area.

The mean concentration of the trace metals in muscles of *Anabus testudineus* collected from the sampling site is given in Table 2. The Mn concentration ranged from (2.659-4.844 ppm) with mean concentration of 3.809 ppm. The Zn concentration ranged from (2.100-3.954 ppm) with mean concentration at 3.712 ppm. The concentration of Cu ranged from (0.048-0.412 ppm) with a mean concentration of 0.211 ppm. For Pb, the concentration ranged from (0.19-0.238 ppm) with mean value at 0.204. The results showed that Zn and Mn in fish accumulated in significantly higher concentrations compared to copper and lead. In fish samples, the trace metal was recorded to be in the decreasing sequence of Mn>Zn>Cu>Pb, similar to the result found in the water samples. However the results showed that for each element analysed, the trace metal concentration in fish was much higher than that in water (Table 2). Fish, as tertiary consumers in the food chain accumulate these trace metals from the surrounding environment. This indicated the biomagnifications of the trace elements in the food chain. The symptoms of acute lead poisoning are headache, irritability, abdominal pain and various symptoms related to the nervous system (Jarup L, 2003). Effects of Mn are decrease in systolic blood pressure, disturbed excretion of 17- ketosteroids, change in erythropoesis and granulocyte formation. Effects of high dose of copper leads to severe mucosal irritation and corrosion, wide spread capillary damage, hepatic and renal damage and central nervous system irritation followed by depression (WHO, 1993).

Table 3. Concentration of trace elements in water samples of Deepor Beel as compared to CPCB (1986) and WHO (1993)

Trace element	Range	CPCB (surface water)	WHO (drinkingwater)
Manganese	0.5149-1.5314	2	0.05
Zinc	0.1712-0.9472	5	0.05
Copper	0.008-0.312	3	2
Lead	0.01-0.118	0.1	0.01

Fishing activity is a major source of livelihood of the people residing in the neighbouring villages around Deepor beel and fishes captured in Deepor beel are sold to consumers in the local market. Though study related to the affect of the six years old dumping ground on the neighbouring villagers is yet to be done it is evident from the high level of these selected elements in fish tissue will be a major health hazard to the consumers. The low level of these elements in control fish samples reveal that biomagnifications of these trace elements take place in subsequent trophic levels in contaminated areas.

Conclusion

The results obtained in our pilot scale study though preliminary it clearly shows that the MSW dumping ground contiguous to the beel has played a major role in deteriorating the water quality status of Deepor beel. Deepor beel harbours a rich biodiversity in terms of migratory birds and fish is the major food for these visiting birds. Long term unsorted dumping of MSW has not only endangered the water resources in its vicinity but also caused health hazard to the people residing in the surrounding villages and possible biomagnifications of toxic elements in the body of the fauna present in it.

References

APHA *Standard method for examination of water and wastewater* (20th ed.) Washington, DC: APHA (American Public Health Association) . 1998, 3-5-3-8.

Eneji I S, Sha'Ato R and Annune P A, An assessment of heavy metals loading in river Benue in the Makurdi Metropolitan area in Central Nigeria, *Environ Monit Assess*, 184 (2012) 201-207.

Jorstad L B, Jankowski J and Acwrth R I, Analysis of the distribution of inorganic constituents in a landfill leachate-contaminated aquifer Astolabe Park, Sydney, Australia, *Environmental Geology*, 46 (2004) 263-272.

Jarup L, Hazards of heavy metal contamination, *British Medical Bulletin*, 68 (2003) 167-182.

Kosygin L, Dharmendra H and Gyaneshwari R, Pollution status and conservation strategies of Moirang river, Manipur with a note on its aquatic bio-resources, *Journal of Environmental Biology*, 28(3) (2007) 669-673.

Mohiuddin K M, Otomo K, Ogawa Y and Shikazono N, Sseasonal and spatial distribution of trace elements in the water and sediments of the Tsurumi river in Japan, *Environ Monit Assess*, 184 (2012)

Planning commission, Planning commission report for Deepor beel, planningcommission.nic.in/reports/E F/Deeporbeel.pdf 2008.

Roy S and Kalita J C, Identification of estrogenic heavy metals in water bodies around Guwahati city, Assam, India, *International Journal of Chemtech Research*, 3(2) (2011) 699-702.

Ramos L, Fernandez M A, Gonzalez M J and Hernandez L.M, Heavy metal pollution in water, sediments, and earthworms from the Ebro river, Spain, *Bull. Environ. Contam. Toxicol*, 63 (1999) 305.

WHO, WHO standards for drinking water, water>drinking water standards 1993.

Session- IV

Paleo Climatic Conditions

Cedrus deodara (Roxb.) G. Don a possible indicator of climate change in Eastern Himalaya

Prabha Sharma & P. L. Uniyal*

Department of Botany, University of Delhi, Delhi – 110 007, India
[E-mail: sharmaprabha3@gmail.com, uniyalpl@rediffmail.com]

Introduction

The Himalaya is home to the world's highest peaks, consisting of about 50 peaks which are more than 7500 meters high above mean sea level. The Himalaya shape the majority of snow covered region outside the polar region and the Himalayan glaciers are extremely susceptible to the on-going warming. The detailed glacier inventory of Indian Himalayas and details about glaciers in the districts of Uttrakhand, Himachal Pradesh and Jammu and Kashmir States give a clear picture of global warming and climate change (GSI, 2009). Various technologies like remote sensing techniques attempted to understand the changes in the Himalayan Cryosphere. The studies on the changes in glacial amount, glacial mass balance and cyclic snow cover; found the glacial retreat for 1868 glaciers in 11 basins since 1962 at 16% (overall deglaciation). The observation of the seasonal snow cover in 28 river sub-basins in central and western Himalayas also pointed snow retreat even throughout winters (Kulkarni *et al.*, 2011). Changes in snow cover in the Hindu kush-Himalayan region using MODIS data from 2000 to 2010 have also been examined (Gurung *et al.*, 2011). The climate above the mountain area of the Western Himalaya is exceedingly changeable. It ranges from subtropical, wet-temperate and moist temperate to cold-arid. The region consists of snow covered mountain, glaciated valleys and a network of rivers. The typical geographical location of the Himalaya from west to east, the distinction in heat and rain clearly obvious in the vegetation and glacier margins and the time and degree of snow melt discharge into rivers.

Materials and methods

Changes in the past or present in the climate at global scale may be left their enduring route over the region, although it is well undestood that the overall change in climate over the region is the sum of the multifaceted interactive physical system operating at world, region and local level highlighting a warming trend over the Western Himalaya in 20th century observed in many other parts of the globe and also over the Indian region (Pant *et al.*, 2003, Bhutiyani *et al.*, 2006, Jones *et al.*, 1999, IPCC, 2001, Jones *et al.* 2003). However, we must clearly ask the questions about recent warming patterns in Himalaya, Little Ice Age cooling and Medieval period warming addressing to regional high resolution alternative data such as tree-ring chronologies.

Results and discussion

Tree rings studies in India have been documented since long back (Gamble, 1902), with limited applications related to forestry aspects relating to evaluation of growth rates, wood productivity and quality, or rotation cycles. Tree rings gained importance for the reconstruction of various aspects of climate and other environmental issues very late. The exceptional climate of India is by power of seasonal wind patterns and linked changes of precipitation and temperature. The wealth of the nation is regularly reliant on vagaries of the monsoon, but the active evidence of the monsoon is not extensive enough to make consistent models for climatic forecast. Thus it is indeed to understand the long-standing facets of the Indian climate. In this circumstance, tree-ring based climate renovation can have a key role in providing high-resolution account that envelop the last several centuries depending on the accessibility of samples from older trees (Chowdhury, 1940).

Organized tree-ring investigation relating precise dating of lengthy series of growth rings had merely been initiated since the finish of the 1980s. After that extensive amounts of research efforts on tree-ring analyses connected to climate have been done for the Himalayan and peninsular region of the Indian subcontinent. Most trees are confirmed to have datable tree rings but only few have been used for climate rebuilding and additional feature, *e.g.*, glacial variation or palaeo-seismic dating. Henceforth, future studies on other tree-ring parameters required to be analyzed for a superior understanding of the regional climate and its association with other climatic happening in a universal viewpoint (Bhattacharyya & Shah, 2009). Tree-ring chronologies, while relatively tiny in palaeoclimatic time frame, are accurately dated and the time resolution is yearly or seasonal. Tree-ring examination of conifers (*Pinus, Picea, Cedrus, Abies*) from the Western Himalayan region point to many tree species in the area more than 500 years in age with clearly marked and environmentally receptive annual growth rings providing yearly proxy facts to learn climatic and hydrologic disparity over both space and time (Bhattacharyya *et al.,* 1988, Hughes *et al.,* 1992, Borgaonkar *et al.*, 1999, Singh *et al.,* 2005, Singh *et al.,* 2006).

Cedrus deodara

Cedrus deodara (Roxb.) G. Don, is ecologically and economically highly valuable taxa belonging to the unique and diverse Gymnosperms (Fig. 1). The plant is significant through its various uses, being carrying of chemical compounds and resins, which are great interest to diverse religion, culture and people such as foresters, silviculturists, ecologists, environmentalists, and biologists. The wood extract is carminative, diaphoretic and antipyretic, used to treat flatulence, pulmonary and urinary disorders, rheumatism, piles and kidney stones, also has antispasmodic and spasmolytic, dermal application mast cell-stabilizing and lipoxygenase-inhibitory, immunological activities. The root extract or oil, also contains terpenoids such as himachalol, atlantone and trans-atlantone used as an anti-ulcer drug by Hakims. Deodar has major application in wood industry, mainly used in buildings for beams, door and window frames and shutters, furniture and cabinet making, for overhead power and telecommunication lines for manufacture of high class pencils, cooling towers, drawing equipments, jute and textile mill accessories, railway carriages and model (Dirr, 1990).

Fig. 1. *Cedrus deodara* growing in North Sikkim Himalaya depicting tree ring, indicator of climate change

According to the current conservation status it belongs to threatened species which is least concerned and at lower risk whose IUCN conservation status need updating. The future oriented, successful conservation strategies of threatened species are impossible without knowledge of its biology, taxonomy, morphological and genetic variability, distribution, conservation status and the necessary first step in this direction is identification of the organism. The categorization, description of species, their interdependence and distribution across the globe are essential components of biodiversity.

Sikkim

Sikkim is a small hilly state in the Eastern Himalaya. Sikkim is endowed with a rich biodiversity of flora and fauna. Sikkim is at top with respect to the number of flowering plants (26%), around 4500 species or number of endemic and threatened species (65 species). Conservation of these species and their habitat warrants special attention in the State, especially in the context of climate change. Also climate change posing a threat to the production and productivity of the cash crops and livestock products in Sikkim, which are significant from the point of view of the livelihoods they provides to the people (SAPCC, 2011). Whatever land is left for cultivation is besieged by variable agro-climatic conditions, difficult terrain, steep slope, acidic soil and prolonged dry spell. In a predominantly rural economy such as Sikkim, animal husbandry activities form an extremely important element in the effort to bring about substantial improvements in living standards.

Vast studies have been done to untie the climatic changes during the recent past in the Himalayan region, mostly in its western and central parts, using tree-ring data (Hughes, *et al.*, 1992, Borgaonkar *et al.*, 1999, Bhattacharyya *et al.*, 1999, Bilharn *et al.*, 1983, Yadav *et al.*, 1992, Yadav *et al.*, 1999). It is essential to evaluate the dendroclimatic potential and problems in the Eastern Himalayan region which is below the pressure of a monsoon climate and for whom climatic records are scarce. Such studies certainly add the knowledge for different phases of climate changes in the complete Himalayan region and to analyse their relationship to supplementary phenomena in a comprehensive context (Chaudhary *et al.*, 1999).

Conclusion

Several conifers growing in the Eastern Himalayan region have been found to have datable tree-ring series and these provide data to prepare climatically sensitive tree-ring chronologies. There are substantial number of unexplored tree species in the Himalaya that are appropriate for diverse facets of tree-ring research to understand the regional climate of India. More samples and extensive collections from undisturbed forest sites in remote areas should be the prerequisite for tree ring analysis. In addition, insufficient meteorological information close to the sampling sites and meager records in the Eastern Himalayan region create problems in analysing tree growth-climate relationships.

Multivariate regression analyses using multiple site chronologies and regional climatic data might be appropriate for the climatic reconstruction of Eastern Himalaya. A set of relations of climatically sensitive tree-ring chronologies from trees growing in distinct climatic regions of the north-western and north-eastern part of the Himalaya would be of great meaning in organizing long-term, high-resolution climate reconstructions which will help in understanding the detailed climatic dynamics of the Indian subcontinent in a global perspective. To achieve millennium long tree-ring records, widespread efforts require to be made in the assemblage of tree-ring samples in unique geographical areas (Bhattacharyya *et al.*, 2009, Chaudhary *et al.*, 1999).

Recent significant warming faced globally, focused more attention of leading academicians and politicians to understand the cause and effective communication on the ecosystems and their individual components to climate changes. The high altitude mountain systems are most sensitive to such climate changes in terms of vegetation shifts, faunal status, and glacier fluctuations. Climate change is a major concern in the Himalayas because of potential impacts on the economy, ecology, and environment of the Himalayas and the areas downstream. The rapid urbanization of the hill stations where the observatories are located may also be one of the factors responsible for the observed increase. In the absence of long period data with a dense network, proxy climatic records including tree-rings provide a very useful tool in understanding the contemporary climatic change patterns over the Himalayas (Singh & Yadav, 2000, Borgaonkar *et al.,* 2010).

Deodar (*Cedrus deodara*) growing in varied natural circumstances has been found to supply ideal tree ring substance for developing incredibly long chronologies in India (Bhattacharyya *et al.*, 2009). Henceforth, systematists and systematic studies including Tree ring analysis through morphological and anatomical studies are significant to contribute to conservation and biodiversity assessment and to make taxonomic data more accessible to a larger audience in the context of attention regarding environment, human health and climate change.

Sikkim Himalaya have environmental, aesthetic, social and economic significance, for the local people for day-to-day subsistence from mountain resources and also for the 40% of the worldwide population depending ultimately on these resources for water, hydroelectricity, timber, biodiversity and niche products (Schild, 2008, Schild & Sharma, 2011). Sikkim is endowed with a rich biodiversity of flora and fauna. Shifts in the altitudinal range of species and communities are a common impact of climate change; however, evidence based on research is scanty. In a recent study on endemic species of the alpine belt in the Sikkim Himalayas (>4,000 masl), report 87% of the 124 endemic plant species are facing pronounced upward shift over the past 150 years (Bajracharya *et al.*, 2011). Conservation of these species and their habitat warrants special attention in the State, especially in the context of climate change (Telwala, 2011). Also, the trees are often cut and sold by local people as a livelihood, demanding constant supply of high quality seed which can only be guaranteed through reforestation practices.

Acknowledgement

Authors are thankful to University of Delhi, The Agharkar Research Institute Pune and Department of Science and Technology, New Delhi Govt. of India for the award of Woman Scientist Fellowship

References

Bajracharya S R, Maharjan S B & Shrestha F, 'Glaciers shrinking in Nepal Himalaya.' In: *Climate change: Geophysical foundations and ecological effects*, edited by J. Blanco, & H. Kheradmand, (Rijeka, Croatia) 2011, pp. 445–458.

Bhattacharyya A & Shah S K, Tree –ring studies in India past appraisal, present status and future prospects. *IAWA J.*, 30 (2009) 361–370.

Bhattacharyya A & Yadav R R, Climatic reconstruction using tree-ring data from tropical and temperate regions of India – A review. *IAWA J.*, 20 (1999) 311–316.

Bhattacharyya A, LaMarche V C & Telewski F W, Dendrochronological reconnai-ssance of the conifers of Northwest India. *Tree-Ring Bull.*, 48 (1988) 21-30.

Bhutiyani M R, Kale V S & Pawar N J, Long-term trends in maximum, minimum and mean annual air temperature across the Northwestern Himalaya during the twentieth century. *Climate Change,* 85 (2006) 159-177.

Bilharn R, Pant G B & Jacoby G C, Dendroclimatic potential of conifer trees from the Sir Sar range in the Karakoram. *Mausam,* 7(1983) 45-50.

Borgaonkar H P, Pant G P & Rupa Kumar K, Tree-ring chronologies from Western Himalaya and their dendroclimatic potential. *IAWA J.,* 20 (1999) 295-309.

Borgaonkar H P, Sikder A B & Somaru R, High altitude forest sensitivity to the recent warming: A tree-ring analysis of conifers from Western Himalaya, India. *Quatern. Int.,* 30 (2010) 1–9.

Chaudhary V, Bhattacharyya A & Yadav R R, Tree-Ring studies in the Himalayan region: Prospects and problems *IAWA J.*, 20 (1999) 317-324.

Chowdhury K A, The formation of growth rings in Indian trees. III. *Indian For. Rec.,* 3 (1940) 59–75.

Dirr, M.A., *Manual of Woody Landscape Plants: Their Identification, Ornamental Characteristics, Culture, Propagation, and Uses* (Stipes Publishing Company, Champaign, IL) 1990, pp. 1, 007.

GSI : *Inventory of Himalayan glaciers* edited by C.V. Sangewar and S.P. Shukla , 2009, pp. 588.

Gurung D R, Kulkarni A V, Giriraj A, Aung K S, Shresstha B & Srinivasan J, Changes in seasonal snow cover in Hindu Kush-Himalayan region. *The Cryosphere Discuss.,* 5 (2011) 755-777.

Gurung D R, Kulkarni A V, Giriraj A, Aung K S, Shresstha B & Srinivasan J, Gamble J S, *A manual of Indian timbers*, (Sampsonlow, Marston and Co., London) 1902.

Hughes M K, Dendroclimatic evidence from the Western Himalaya. In: *Climate sinee AD 1500* edited by R. S. Bradley & P. O. Jones Routledge, London 1992, pp. 415-431.

IPCC, *Climate Change : The Scientific Basis*, Contribution of Working Group-I to the Third Assessment Report of Intergovernmental Panel on Climate Change (IPCC). Cambridge University Press, Cambridge, 2001, pp. 881.

Jones P D & Moberg A, Hemispheric and large scale surface air temperature: an extensive revision and updates to 2001. *J. Climatol.*, 16 (2003) 206–223.

Jones P D, New M, Parker D E, Martin S & Rigor I G, Surface air temperature and its changes over the past 150 years. *Rev. Geophys.*, 37 (1999) 173–199.

Kulkarni A V, Rathore B P, Singh S K & Bahuguna I M, Understanding changes in the Himalayan Cryosphere using remote sensing techniques. *Int. Jour. Remote Sens.*, 32 (2011) 601-615.

Pant G B, Borgaonkar H P & Rupa Kumar K, Climate variability over the western Himalaya since the little ice age: dendroclimatic implications. *Jalvigyan Sameeksha,* 18 (2003) 111–121.

SAPCC *Sikkim Action Plan on Climate Change* (2012-2030), 2011, pp. 1-160.

Schild A & Sharma E, 'Sustainable mountain development revisited.' *Mt. Res. Dev.*, 31(2011) 237–241.

Schild A, 'The case of the Hindu Kush-Himalayas: ICIMOD's position on climate change and mountain systems.' *Mt. Res. Dev.*, 28(2008) 328–331.

Singh J & Yadav R R, Spring precipitation variations over the Western Himalaya, India, since A.D. 1731 as deduced from tree-rings. *J. Geophy. Res.*, 110 (2005) 1-8.

Singh J & Yadav R R, Tree-ring indications of recent glacier fluctuations in Gangotri, Western Himalaya, India. *Curr. Sci.*, 79(2000) 1598–1601.

Singh J, Park W K & Yadav R R, Tree-ring based hydrological records for Western Himalaya, India, since A.D. 1560. *Clim. Dynam.*, 26 (2006) 295–303.

Telwala Y, *Biological response to climate change in Sikkim Himalaya.* PhD thesis, University of Delhi, New Delhi, India 2011.

Yadav R R & Bhatlacharyya A, A 745-year chronology of *Cedrus deodara* from western Himalaya, India and its dendroclimatic potential. *Dendrochronologia,* 10 (1992) 53-61.

Yadav R R, Park, W K & Bhttacharyya A, Spring-temperature variations in Western Himalaya, India as reconstructed from tree-rings: AD 1390–1987. *Holocene,* 9 (1999) 85–90.

Impact of climate change on heterotrophic bacterial communities in the water and sediment of kongsfjord in Norwegian Arctic

Hatha A.A.M.[1], *Mujeeb Rahiman*[1] *K.M., Deepu Lal K.M.*[2], *Krishnan K.P.*[3], *Rupesh Kumar Sinha*[3] *& A.V. Saramma*[1]

[1]Department of Marine Biology, Microbiology and Biochemistry, School of Marine Sciences, Cochin University of Science and Technology, Cochin – 682 016, India.
[2]Department of Chemical Oceanography, School of Marine Sciences, Cochin University of Science and Technology, Cochin – 682 016, India.
[3]Cryobiology Lab, National Centre for Antarctic and Ocean Research, Ministry of Earth Sciences, Headland Sada, Vasco da Gama, India
[E-mail: mohamedhatha@gmail.com]

Introduction

Cold environments represent a large part of the Earth's biosphere, and their microbiota are of increasing interest. Polar regions are of interest since they provide diverse terrestrial and marine habitats for psychrophilic and psychrotrophic microorganisms. The vast numbers of cold-adapted microorganisms which successfully inhabit these regions, are referred to as psychrophilic and psychrotolerant organisms, play vital roles in global elemental cycles and in the mineralization of organic matter (Jiao *et al.* 2010, Piontek, Borchard *et al.* 2012). Unlike deep oceans, polar marine environments are subject to large seasonal variations in sea-ice cover, greatly affecting the biology of organisms (Vadstein, 2011). Currently, the knowledge of polar microorganisms based on ecological and genomic perspectives is in the early phase of an exponential growth (Russo, R. *et al.* 2010, Iversen, K.R. & Seuthe, L., 2011).

The Arctic fjord Kongsfjorden located off the west coast of Spitsbergen have attracted the attention of scientists as a model site for studies on the impact of climate change in the Arctic (Piontek *et al.*, 2011; Iverson and Seuthe, 2012) and they are regarded as key European sites for Arctic biodiversity monitoring[6]. Earlier studies in these fjords have focused both on the physical characteristics of the environment and the biotic components of the ecosystem (Hop, H., Pearson, T., *et al.* 2002, Shivaji S, Kiran M D & Chintalapati S 2007). Despite the fact that Arctic has a number of fjords comparatively little is known about the bacterial diversity of the fjords (Jankowska, K., Wlodarska-Kowalczuk, *et al.* 2005, Teske, A., Durbin, A. 2011).

Considering the extent of warming in the arctic region and the resultant changes in the dynamic marine environments there is a need to monitor the bacterial diversity in the fjord environments, especially in terms of cultivable bacteria. The present study reports the diversity of cultivable heterotrophic bacteria from the water and sediment samples of Kongsfjord, their growth responses to important environmental variables and ability to produce industrially important hydrolytic enzymes.

Materials and methods

Sampling was carried out from 4 stations in the Kongsfjord (Fig. 1), Norwegian Arctic on 4th July 2009. Kongsfjorden (79°58′N, 12°E) is a small fjord with a wide opening to the open ocean via Kongsfjordrenna. A sill in the middle of the fjord divides it in an outer part, strongly influenced by West Spitsbergen Current, and an inner part that is under the impact of 4 different glaciers namely, Kronebreen, Kongsvegen, Conwaybreen and Blomstrandbreen (Svendsen, H., Beszczynska-Möller, *et al.* 2002).

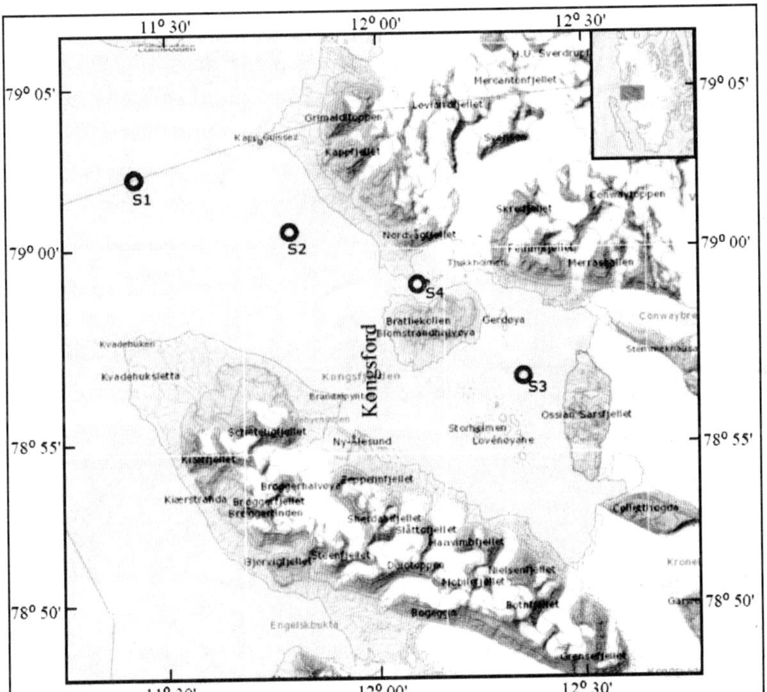

Fig. 1. Location of sampling sites in Kongsfjorden fjord, Ny Alesund, Arctic

The water and sediment samples were collected on board Research Vessel 'Teisten' and position of the sampling site was noted using the global positioning system (GPS, Raymarine E 120) system installed on-board Teisten. Water and sediment samples were collected with the help of Niskin sampler and Van Veen Grab (Model 12.320, KC, Denmark) respectively. Depth of the sampling site, water temperature and salinity were measured using the STD/CTD instrument model SD204 (SAIV A/S Environmental Sensors and Systems, Norway). Soon after hauling up the sample on board, water and sediment temperature was checked by centigrade digital thermometer and salinity by refractometer (Atago, Japan). Water and sediment samples were collected aseptically in sterile Duran Schott glass bottles and polythene bags respectively. Samples were transported to the Kings Bay marine lab at the International Arctic Research Station within 4 hours of collection and processed for retrievable heterotrophic bacteria. Particle size analysis of the sediment samples was done by laser diffraction method using particle size analyzer (SympaTEC, Germany).

Isolation and characterization heterotrophic bacteria

The water and sediment samples were serially diluted and 0.2 ml of various dilutions was spread plated on Zobell marine agar (ZMA). Plates were incubated at 10°C for 1-2 weeks. Plates with 25-250 colonies from ZMA were selected for enumeration and expressed as cfu/ml or g of water and sediment samples respectively. Morphologically different colonies from ZMA were isolated, restreaked to ensure purity and characterized and grouped through biochemical tests (Holt J G, Krieg N R, *et al.* 2000) and protein profiling (Laemmli, U.K., 1970, Bradford, M.M., 1976). The selected isolates from this group were further characterized at molecular level using 16s rRNA gene analysis.

PCR amplification of the 16S rDNA, sequencing and phylogenetic analysis

Total bacterial genomic DNA was extracted using the Bacterial Genomic DNA (prep) Kit (Chromous Biotech, India). DNA extracts were verified by gel electrophoresis and 16S rDNA genes were amplified by using the polymerase chain reaction (PCR) with the universal primers 27f (5'-AGAGTTTGATCC-TGGCTCAG-3') and 1492r (5'- GGTTACCTTGTTACGACTT-3') (Bosshard, P.P. & Santini, Y. *et al.* 2000).

The PCR reaction was set up using 100 ng of genomic DNA, 1.5 μl of 10 pmol of both forward and reverse primers, 5 μl of standard 1X Taq buffer, 3 μl of 1.5 mM $MgCl_2$, 0.4 μl of 200 μM dNTP mix and 0.3 μl of 1.5 U Taq DNA polymerase (Banglore Genei Pvt. Ltd., Bangalore, India) in 50 μl reaction volume. The PCR amplification was carried out in PCR tubes using Biorad MJ Mini personal thermal cycler (Model PTC 1148, USA), for 30 cycles after an initial denaturation at 94°C for 5 min. Each PCR cycle comprised three steps: 30s at 95°C (denaturation), 30s at 45°C (annealing) and 120s at 72°C (extension). A final extension of 10 min was given at 72°C. PCR products were checked by gel electrophoresis and purified using the PCR Clean-Up Kit (Chromous Biotech, India) and sequenced using an ABi 3730 XL Genetic Analyser (Applied Biosystems, USA).

The almost complete 16S rRNA gene sequences (1418-1542 bases) obtained was subjected to BLAST sequence similarity search (http://blast.ncbi.nlm.nih.gov/BLAST) to identify the nearest taxa. 16S rRNA gene sequences were aligned using the CLUSTAL W and phylogenetic trees were constructed using two tree making algorithms, maximum likelihood (ML) and Neighbour-joining (NJ) methods using MEGA version 5 (Tamura, K., Peterson, D., *et al.* 2011). Two consensus trees were finally constructed to place isolates from water and sediment samples.

Nucleotide sequence and accession numbers

All the 16S rRNA gene sequences of the strains selected from each group were deposited in GenBank with accession numbers JX262392 to JX262406.

Result and discussion

Exact geographical co-ordinates of the sampling station, depth, surface and bottom water temperature, sediment temperature and salinity of water are given in Table 1. While surface water temperature ranged from 3.61 (station III) to 6.13 (station I), bottom water temperature was around 1°C. Stations III and IV were much closer to the front moraines of two major glaciers and the lower temperature was anticipated. The sampling was conducted in Arctic summer, during which the oceanography of the fjord undergoes rapid and considerable change. The summer situation suggested an unobstructed exchange of water masses between the open sea, the shelf and the fjord itself (Walkusz, W., Kwasniewski, *et al.* 2009). Kongsfjord was influenced by Atlantic-derived water masses covered by thin layer of freshwater runoff from the glaciers. Due to the great discharge of freshwater from the glaciers, the stations closer to them shows strong stratification in temperature and salinity with the advancement of summer season. However, this gradient diminishes towards the fjord opening.

Table 1. Geographical location of sampling sites and physiochemical properties of water and sediment samples from Kongsfjord during summer 2009.

Sampling stations	Location	Depth of the station (m)	Surface Water temp. (°C)	Bottom Water temp. (°C)	Water salinity (ppt)	Sediment temp (°C)
Station I	79°02'36" N 11°19'14" E	320	6.13	0.8	33.69	1.0
Station II	79°01'22" N 11°43'06" E	185	6.51	2.3	33.56	1.5
Station III	78°57'29" N 12°19'27" E	45	4.14	0.8	33.13	-0.6
Station IV	79°00'01" N 12°02'50" E	110	3.61	1.2	33.11	-0.8

Due to the large scale land run-off of the melt water the inner basins of the Kongsfjord were found to be reddish brown, indicating massive influx of terrogenous sediments. The outer basins and the sampling station I at fjord opening were characterized by clear blue waters. Similar observations were reported previously at (Kongsfjorden Zajaczkowski, 2008, Piwosz, Walkusz *et al.* 2009).

Grain size analysis of the sediment samples revealed silty-clay nature of the sediments from all stations, except that from station III, which was of silty sand (Fig. 2). While silt fraction ranged from 67% at station III to 82% at station I, clay fraction ranged from 12% (station II) to 23% (station IV). Fairly large proportion of sand was noticed in the sediments of station III. As this station is closer to the front moraine of glacier, the sand transport mediated through the glacier might have contributed to this fairly significant sand fraction in the sediments at this station. The sediments from all the four stations were visually distinguishable in their color which varied from light brown to dark chocolate. The color variation may be due to variation in the organic content of the sediment and oxygenation at the bottom (Jorgensen *et al.* 2005).

Retrievable bacterial count

Retrievable bacterial load of four water and sediment samples were ranged from 1.22×10^3 to 1.68×10^3 cfu/ ml and 1.35×10^5 to 8.45×10^6 cfu/g respectively. The sediment samples from station III (inner to the fjord) showed highest retrievable count of bacteria than other stations. As expected more heterotrophic bacteria were retrieved from the sediment samples, which were nearly 2 log higher than the load encountered in the water samples. In general sediments act as a repository of bacteria and offer much more stable and protective environment for the survival of them. To the contrary, the conditions in the water column are highly dynamic, especially in the summer when there is a large scale influx of melt water, which can seriously alter the salinity, at least at the surface water, and pose stress and injury to the microbial community. Such injured cells, though viable, might not develop on the media, which in general have some inhibitory substances. Though the land run-off during the summer result in large scale influx of terrogenous organic carbon, which are more or less recalcitrant and offer little help by way of providing nutrition (Kirchman *et al.* 2009). The retrievable count encountered in the sediment samples were higher than those reported (Srinivas *et al.* 2009). Relatively higher density of cultivable bacteria were reported from inner shelf sediments (Zheng *et al.* 2009), which is similar to our observations.

Benthic bacterial communities in the ocean environment play a significant role in ecological and global biogeochemical cycle, because they can rapidly degrade and utilize particulate organic matter (Teske *et al.* 2011, Bowman *et al.* 2003, Michaud *et al.* 2004) and regulating the transformation of biogenic elements such as C, N, P, Fe, O, and S (Kirchman *et al.* 2009, Vandieken *et al.* 2006, Selje *et al.* 2004). Microbial community structure analysis is important for an understanding of ecosystem processes and in defining the roles that bacteria play in overall oceanic processes and bacterioplankton components is related to oceanic water masses and controlled by their environmental and biogeochemical properties (Cottrell *et al.* 2000). It is important to know which phylogenetic groups of bacteria dominate marine bacterioplankton communities because abundant groups may be proportionally more influential in carbon cycling and other biogeochemical processes (Zengler *et al.* 2009). In culture-dependent methods, the bacteria obtained allow for detailed studies and provide more information about the physiological and metabolic characteristics of bacteria, and microbial response to environment change etc (Zeng *et al.* 2011, Denton *et al.* 1998).

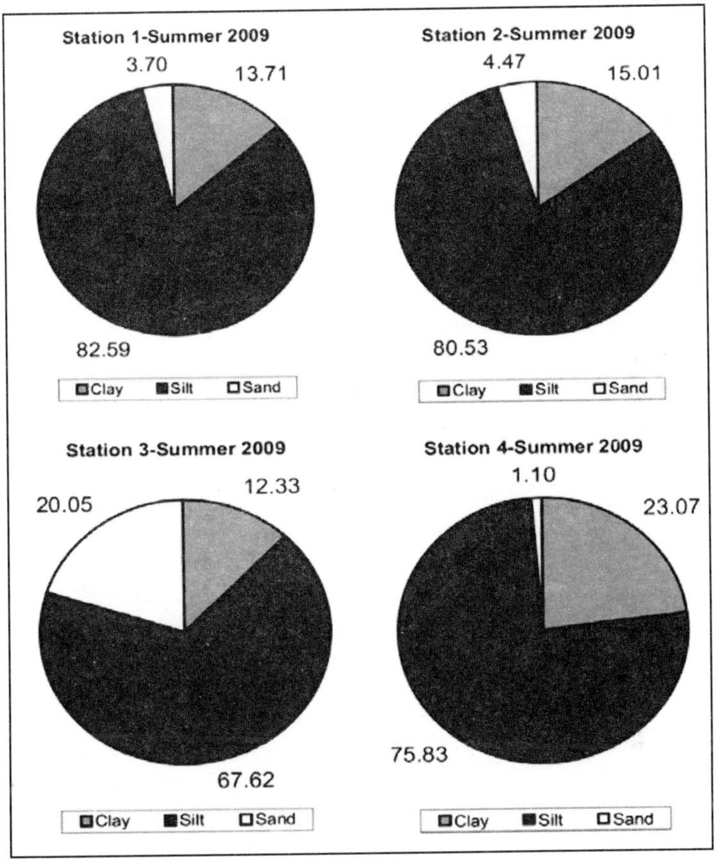

Fig. 2. Percentage particle size analysis of sediment samples from Arctic Ocean during summer 2009

Major groups of bacteria

A total of 298 bacterial strains were characterised from water and sediment samples and based on the biochemical test and protein profiing, the isolates could be categorised into 33 groups. The nearest phylogenetic neighbour of the selected 33 isolated strains from each group were identified following BLAST analysis of the 16S rRNA gene sequence. Sequence similarity of the strains compared to the nearest phylogenetic neighbour ranged from 97 to 100% (Table 2). Based on 16S rRNA gene sequence similarity, the isolates were categorised into 8 phylotypes belonging to two phyla (Fig. 3) groups from water samples. Gammaproteobacteria and Bacilli were the two bacterial phyla identified from the surface water samples of Kongsfjord in the present study (Fig. 3). Gammaproteobacteria was the major group with more than 80% of occurrence. The Gammaproteobacteria group also showed good diversity with various heterotrophic bacteria such as *Enterobacter ludwigi, Enterobacter cancerrogenous, Halomonas boliviensis, Pseudomonas sabulinigri, Pseudomonas fragi, Pseudomonas koreensis* and *Stenotrophomonas maltophila. Stenotrophomonas* strains formed major share of the isolated Gammaproteobacteria which raises some concern as this organism is increasingly being recognized as an important cause of nosocomial infection (Kwa *et al.* 2008, Paez *et al.* 2008) and severe infections caused by *S. maltophila* are associated with high mortality (Orcutt *et al.* 2011). It is considered as an uncommon bacterium in the arctic environment, though there are reports of its occurrence in the marine realm (Zheng *et al.* 2009, Zeng *et al.* 2011). It would be interesting to study whether the fast retreating ice cover of the Arctic Ocean and resultant opening up of sea routes as well as mixing of water masses from the lower latitudes playing a role in bringing more mesophilic fauna into higher latitudes. *Pseudomonas fragi* is a typical psychrophilic bacterium, which unlike most other members of the *Pseudomonas* genus, does not produce siderophores. This is an indirect indication that the Kongsfjorden water does not pose any nutrient limitation, in terms of iron. Dominance of Gammaproteobaceria and presence of Firmicutes were reported from Kongsfjorden (Zheng *et al.* 2009) and northern Bering Sea (Perreault *et al.* 2007). *Bacillus thuringiensis* was the sole Firmicute encountered in the water samples collected from the Kongsfjord. *Bacillus* was previously reported from sea water and spring samples of Arctic Ocean (Perreault *et al.* 2008, Zeng *et al.* 2004, Srinivas *et al.* 2009).

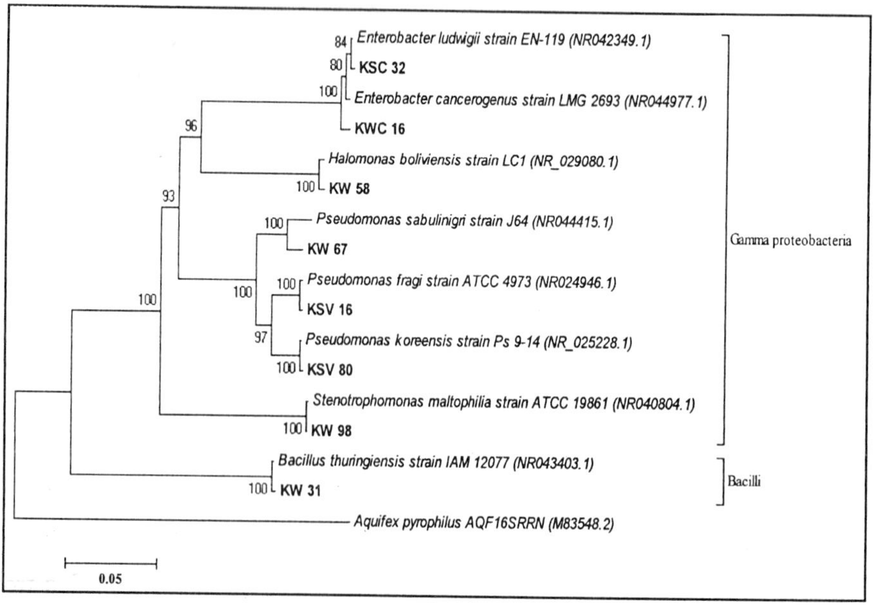

Fig. 3. Phylogenetic trees based on 16S rRNA gene sequences showing the relationship of bacterial strains obtained from water samples collected from Kongsfjord, Arctic, with their nearest phylogenetic relatives

Predominance of Gammaproteobacteria group has been reported from the Arctic fjords of Spitzbergen and from spring samples of Arctic was reported previously (Fogg 1986, Groudieva *et al.* 2004). Reported five different ARDRA profiled *Pseudomonas* from sea ice and sea water from fjords of Spitzbergen, Arctic Ocean. These reports are in agreement with present findings and we have isolated three different species of *Pseudomonas* namely *P. sabulinigri, P. fragi* and *P. koreensis*. Previous studies (Li *et al.* 1999, Yu *et al.* 2006) support the idea that *Pseudomonas* is one of the principal genera of bacteria present in the Arctic Ocean. We did not encounter any Actinobacteria, though the sequences affiliated with the Actinobacteria were detected both in surface and bottom water in Kongsfjorden (Zheng *et. al.* 2009). Three species of *Pseudomonas* (*P. fragi, P. koreensis and P. sabulinigri*) and two species of *Bacillus* (*B. thuringiensis and B. flexus*) encountered in this study were not reported previously from the Kongsfjord. Interestingly we could not find any *Cytophaga – Flavobacterium* group, the isolation of which was reported in some of the previous studies (Reddy *et al.* 2009) from the Arctic fjord environments.

In contrast to the water samples, the sediments samples had good share of Bacilli and Actinobacteria, when compared to Gammaproteobacteria. Gammaproteobacterial presence was restricted to *Stenotrophomonas maltophila* and *Enterobacter cancerogenous*. These two strains of Gammaproteobacteria were also detected in the water samples. Seven phylotypes belonging to three phyla of bacteria were identified from sediment samples (Fig. 4). While Bacilli (42.8%) dominated the sediment samples, *Actinobacteria* and Gammaproteobacteria were equal (28.6%) (Fig. 4). Culture independent methods also reported predominance of Gammaproteobacteria from the cold saline sediment samples of the Canadian high Arctic (Perreault *et al.* 2008) Actinobacteria phyla included phylotypes such as *Brachybacterium paraconglomeratum* and *Micrococcus luteus*, while phylotypes of Firmicutes included *Bacillus thuringiensis, Bacillus flexus* and *Staphylococcus cohnii*. Comprehensive molecular techniques and surveys have revealed that the Actinobacteria account for a small proportion in marine microbial communities, but play a significant ecological role and are ubiquitous component of marine microbial communities. Occurrence of Actinobacteria, Bacilli and Gammaproteobacteria phyla were reported from the fjord sediments of Arctic. However, phylotypes showed variations which could be expected in a highly dynamic environment such as marine environment. The phylotypes of Gammaproteobacteria encountered in the present study included *Stenotrophomonas maltophila* and *Enterobacter ludwigii*. Considerable share of *Stenotrophomonas maltophila* calls for further studies on its virulence and antibiotic resistance properties as it is fast emerging as a bacteria of public health significance. We are in the process of microbial source tracking studies to look at the molecular markers in this organism. Since most of the species identified from water and sediment samples of present study are not reported earlier from Kongsfjorden, present findings give additional knowledge in culturable bacterial diversity of Kongsfjord.

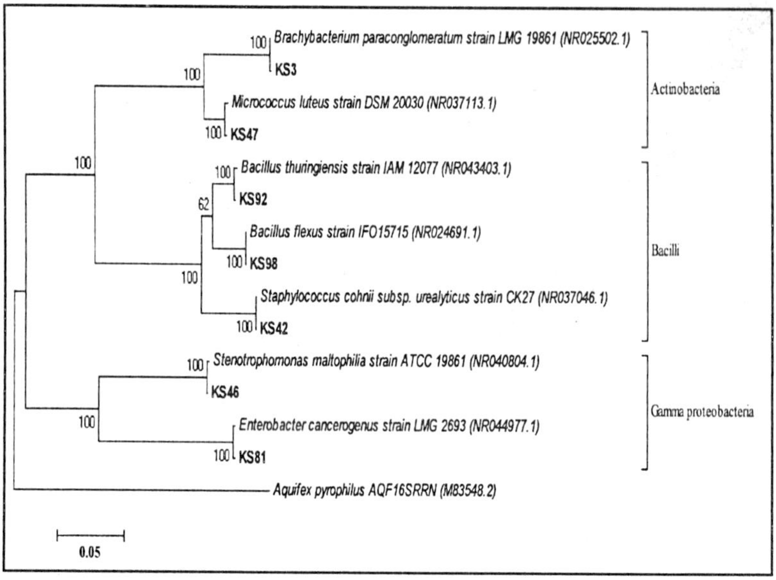

Fig. 4. Phylogenetic trees based on 16S rRNA gene sequences showing the relationship of bacterial strains obtained from sediment samples collected from Kongsfjord, Arctic, with their nearest phylogenetic relatives.

Two *Bacillus* sps. and one *Staphylococcus* sps were identified as Firmicutes from sediment samples (Table 2). Actinobacteria represented by *Brachybacterium* and *Micrococcus* sps. and Gammaproteobacteria by *Stenotrophomonas* and *Enterococcus* (Figure 4). *Bacillus* was also identified from spring samples of Arctic (Perreault *et al.* 2008, Groudieva *et al.* 2004) and melt water stream sediment of Arctic glacier. Though *Staphylococcus* from the spring samples of high Arctic (Groudieva *et al.* 2004), *Stenotrophomonas maltophilia* was not reported from the Arctic Ocean environment.

Table 2. Identification of the bacterial strains isolated from water and sediment samples of Kongsfjord during summer 2009 based on BLAST analysis of the 16S rRNA gene sequences.

S N.	Strain number	Gene accession number	Nearest phylogenetic relative	16S rRNA gene sequence similarity (%)
1	KWC 3, **KWC 16***, KSC 17	JX262394	*Enterobacter cancerogenus* LMG 2693 (NR044977)	98.59
2	KSC 32, KSV 1, KSV 33, KW 17, KW 42, **KW 93**	JX262395	*Enterobacter ludwigii* EN-119 (NR042349)	99.64
3	**KSV 16**	JX262396	*Pseudomonas fragi* strain ATCC 4973 (NR024946.1)	99.71
4	**KSV 80**	JX262397	*Pseudomonas koreensis* strain Ps 9-14 (NR025228)	99.85

5	KW 27, KW 43, KW 44, KW 45, **KW 98**	JX262398	*Stenotrophomonas maltophilia* strain ATCC 19861 (NR040804)	99.78
6	**KW 31**	JX262401	*Bacillus thuringiensis* strain IAM 12077 (NR043403)	99.43
7	**KW 58**	JX262399	*Halomonas boliviensis* strain LC1 (NR029080)	99.43
8	**KW 67**	JX262400	*Pseudomonas sabulinigri* strain J64 (NR044415)	97.7
9	KS 1, KS 7, KS 15, KS 27, **KS 46**	JX262392	*Stenotrophomonas maltophilia* strain ATCC 19861 (NR040804)	99.71
10	**KS 3**	JX262402	*Brachybacterium paraconglomeratum* strain LMG 19861 (NR025502)	99.78
11	**KS 42**	JX262403	*Staphylococcus cohnii* subsp. urealyticus CK27 (NR037046)	100.00
12	**KS 47**	JX262404	*Micrococcus luteus* strain DSM 20030 (NR037113)	99.63
13	KS 51, **KS 92**, KS 96, KS 101	JX262405	*Bacillus thuringiensis* strain IAM 12077 (NR043403)	99.50
14	**KS 81**	JX262393	*Enterobacter cancerogenus* strain LMG 2693 (NR044977)	99.32
15	**KS 98**	JX262406	*Bacillus flexus strain* IFO15715 (NR024691)	100.00

*Bold strain numbers are selected for sequencing

Conclusion

Retreivable heterotrophic bacteria from the water and sediment samples of Kongsfjord were dominated by psychrotolerant strains, indicative of more warm water intrusion into the Kongsfjord. Some species of bacteria identified from Kongsfjord are not reported from earlier; especially three species of *Pseudomonas* and two species of *Bacillus*. Relatively high level of isolation of *Stenotrophomnas maltophila* calls for further studies on this organism, especially in terms of source tracking as well as related to virulence and antibiotic resistance.

Acknowledgements

Authours thank NCAOR and the Ministry of Earth Sciences, Government of India, for providing financial and logistic support for the trip to the Arctic.

References

Bradford M M, Rapid and sensitive method for the quantitation of microgram quantities of protein utilizing the principle of protein-dye binding, *Ana. Biochem.*, 72(1976) 248-254.

Bosshard P P, Santini Y, Gruter D, Stettler R & Bachofen R, Bacterial diversity and community composition in the chemocline of the meromictic alpine lake Cadagnoa reveal by 16S rDNA analysis, *FEMS Microbiol. Ecol.,* 31(2000) 173–182

Bowman J P, McCammon S A, Gibson J A E, Robertson L & Nichols P D, Prokaryotic metabolic activity and community structure in Antarctic continental shelf sediment, *Appl. Environ. Microbiol.,* 69(2003) 2448–2462

Cottrell M T & Kirchman D L, Community composition of marine bacterioplankton determined by 16S rDNA gene clone libraries and fluorescence in situ hybridization, *Appl. Environ. Microbiol.,* 66(2000) 5116–5122

Denton M & Kerr K G, Microbiological and clinical aspects of infection associated with *Stenotrophomonas maltophilia, Clin. Microbiol. Rev.,* 11(1) (1998) 57-80

Fogg G E, *Picoplankton.* (Proceedings of the Royal Society of London Series B 228) 1986, pp. 1-30

Groudieva T, Kambourova M, Yusef H, Royter M, Grote R, Trinks H & Antranikian G, Diversity and cold-active hydrolytic enzymes of culturable bacteria associated with Arctic sea ice, Spitzbergen. *Extremophiles,* 8(2004) 475–488

Hop H, Pearson T, Hegseth E N, Kovacs K M, Wiencke C, Kwasniewski S, Eiane K, Mehlum F, Gulliksen B, Wlodarska-Kowalczuk M, Lydersen C, Weslawski J M, Cochrane S, Gabrielsen G W, Leakey R J G, Lonne O J, Zajaczkowski M, Falk-Petersen S, Kendall M, Wängberg S Å, Bischof K, Voronkov A Y, Kovaltchouk N A, Wiktor J, Poltermann M, di Prisco G, Papucci C & Gerland S, The marine ecosystem of Kongsfjorden, Svalbard, *Polar Res.,* 21(2002) 167–208

Holt J G, Krieg N R, Sneath P H A, Staley J T & Willliams S T, *Bergy's manual of determinative Bacteriology, 9th Edn..* (Lipponcott Williams and Wilkins) 2000 pp. 787

Iversen K R & Seuthe L, Seasonal microbial processes in a high-latitude fjord (Kongsfjorden, Svalbard): I. Heterotrophic bacteria, picoplankton and nanoflagellates, *Polar Biol.,* 34(2011) 731-749

Jiao N, Herndl G J, Hansell D A, Benner R, Kattner G, Wilhelm S W, Kirchman D L, Weinbauer M G, Luo T & Chen F, Microbial production of recalcitrant dissolved organic matter: long-term carbon storage in the global ocean, *Nat. Rev. Microbiol.,* 8(2010) 593–599

Jankowska K, Wlodarska-Kowalczuk M & Wieczorek P, Abundance and biomass of bacteria in two Arctic glacial fjords, *Pol. Polar Res.,* 26(2005) 77-84

Jorgensen B B, Glud R N & Holby O, Oxygen distribution and bioirrigation in Arctic fjord sediments (Svalbard, Barents Sea), *Mar. Ecol. Prog. Ser.,* 292(2005) 85–95

Kirchman D L, Moran X A G & Ducklow H, Microbial growth in polar oceans – role of temperature and potential impact of climate change. *Nat. Rev. Microbiol.,* 7 (2009) 451-459

Kwa A L, Low J G, Lim T P, Leow P C, Kurup A & Tam V H, Independent predictors for mortality in patients with positive *Stenotrophomonas maltophilia* cultures, *Ann. Acad. Med. Singapore,* 37(10) (2008) 826-830

Laemmli U K, Cleavage of structural proteins during the assembly of the head of bacteriophage T4, *Nature (London),* 227(1970) 680-685

Li L, Kato C & Horikoshi K, Bacterial diversity in deep-sea sediments from different depths, *Biodivers. Conserv.*, 8(1999) 659–677

Michaud L, Di Cello F, Brilli M, Fani R, Lo Giudice A & Bruni V, Biodiversity of cultivable psychrotrophic marine bacteria isolated from Terra Nova Bay (Ross Sea, Antarctica), *FEMS Microbiol. Lett.*, 230(2004) 63–71

Orcutt B N, Sylvan J B, Knab N J & Edwards K J, Microbial Ecology of the Dark Ocean above, at, and below the Seafloor, *Microbiol. Mol. Biol. Rev.*, 75(2011) 361–422

Piwosz K, Walkusz W, Hapter R, Wieczorek P, Hop H & Wiktor J, Comparison of productivity and phytoplankton in a warm (Kongsfjorden) and a cold (Hornsund) Spitsbergen fjord in mid-summer 2002, *Polar Biol.*, 32(2009) 549–559

Piontek J, Borchard C, Sperling M, Schulz K G, Riebesell U & Engel A, Response of bacterioplankton activity in an Arctic fjord system to elevated pCO_2: results from a mesocosm perturbation study, *Biogeosci. Discuss.*, 9(2012) 10467-10511

Paez J I, Tengan F M, Barone A A, Levin A S & Costa S F, Factors associated with mortality in patients with bloodstream infection and pneumonia due to *Stenotrophomonas maltophilia, Eur. J. Clin. Microbiol. Infect. Dis.*, 27(10) (2008) 901-906

Perreault N N, Andersen D T, Pollard W H, Greer C W & Whyte LG, Characterization of the prokaryotic diversity in cold saline perennial springs of the Canadian high Arctic, *Appl. Environ. Microbiol.*, 73 (2007) 1532–1543

Perreault N N, Greer C W, Andersen D T, Tille S, Couloume G L, Lollar B S & Whyte L G, Heterotrophic and Autotrophic Microbial Populations in Cold Perennial Springs of the High Arctic, *Appl. Environ. Microbiol.*, 74 (2008) 898–6907

Russo R, Giordano D, Riccio A & di Prisco D, Cinzia Verde Cold-adapted bacteria and the globin case study in the Antarctic bacterium *Pseudoalteromonas haloplanktis* TAC125, *Mar. Genomics,* 3 (2010) 125–131

Reddy P V V, Nageswara Rao S S S, Pratibha M S, Sailaja B, Kavya B, Manorama R R, Singh S M, Srinivas T N R & Shivaji S, Bacterial diversity and bioprospecting for cold-active enzymes from culturable bacteria associated with sediment from a melt water stream of Midtre Lovenbreen glacier, an Arctic glacier, *Res. Microbiol.*, 160(2009) 538-54

Shivaji S, Kiran M D & Chintalapati S, Perception and transduction of low temperature in bacteria, in: *Physiology and biochemistry of extremophiles* edited by C. Gerday & N. Glansdorff, (Washington, DC, ASM Press) 2007 pp. 194–207

Svendsen H, Beszczynska-Möller A, Hagen J O, Lefauconnier B, Tverberg V, Gerland S, Orbaek J B, Bischof K, Papucci C, Zajaczkowski M, Azzolini R, Bruland O, Wiencke C, Winther J G & Dallmann W, The physical environment of Kongsfjorden-Krossfjorden, an Arctic fjord system in Svalbard, *Polar Res.*, 21(2002) 133–166

Srinivas T N R, Nageswara Rao S S S, Reddy P V V, Pratibha M S, Sailaja B, Kavya B, Kishore K H, Begum Z, Singh S M & Shivaji S, Bacterial diversity and bioprospecting for cold-active lipases, amylases and proteases, from culturable bacteria of Kongsfjorden and Ny-Alesund, Svalbard, Arctic, *Curr. Microbiol.*, 59(2009) 537–547

Selje N, Simon M & Brinkhoff T, A newly discovered *Roseobacter* cluster in temperate and polar oceans, *Nature* 427(2004) 445-448.

Teske A, Durbin A, Ziervogel K, Cox C & Arnosti C, Microbial community composition and function in permanently cold seawater and sediments from an Arctic Fjord of Svalbard, *Appl. Environ. Microbiol.,* 77(6)(2011) 2008–2018

Tamura K, Peterson D, Peterson N, Stecher G, Nei M & Kumar S, MEGA5: Molecular evolutionary genetics analysis using maximum likelihood, evolutionary distance, and maximum parsimony methods, *Molecular Biol. Evolution,* 28(10) (2011) 2731–2739.

Vadstein O, Large variation in growth-limiting factors for marine heterotrophic bacteria in the Arctic waters of Spitsbergen (78°N), *Aquat. Microb. Ecol.,* 63(2011) 289–297

Vandieken V, Finke N & Jorgensen, B B, Pathways of carbon oxidation in an Arctic fjord sediment (Svalbard) and isolation of psychrophilic and psychrotolerant Fe(III)-reducing bacteria, *Mar. Ecol. Prog. Ser.,* 322(2006) 29–41

Yu Y, Li H R, Chen B, Zeng Y X & He J F, Phylogenetic diversity and cold-adaptive hydrolytic enzymes of culturable psychrophilic bacteria associated with sea ice from high latitude ocean, Artic, *Wei Sheng Wu Xue Bao (Article in Chinese),* 46(2)(2006) 184-90

Walkusz W, Kwasniewski S, Falk-Petersen S, Hop H, Tverberg V, Wieczorek P & Weslawski J M, Seasonal and spatial changes in the zooplankton community in Kongsfjorden, Svalbard, *Polar Res.,* 28(2009) 254-281

Zajaczkowski M, Sediment supply and fluxes in glacial and outwash fjords, Kongsfjorden and Adventfjorden, Svalbard, *Pol. Polar Res.,* 1(2008) 59–72

Zheng Y, Zheng T & Li H, Community composition of the marine bacterioplankton in Kongsfjorden (Spitsbergen) as revealed by 16S rRNA gene analysis, *Polar Biol.,* 32(2009) 1447–1460

Zengler K, Central role of the cell in microbial ecology, *Microbiol. Mol. Biol. Rev.,* 73(2009) 712–729

Zeng Y, Zou Y, Chen B, Grebmeier J M, Li H, Yu Y & Zheng T, Phylogenetic diversity of sediment bacteria in the northern Bering Sea, *Polar Biol.,* 34(2011) 907–919

Zeng Y, Zou Y, Grebmeier J M, He J & Zheng T, Culture-independent and -dependent methods to investigate the diversity of planktonic bacteria in the northern Bering Sea, *Polar Biol.,* 35(2012) 117–129.

Benthic faunal assemblage - Diversity and time scale changes in the ice cold Kongsfjorden Arctic ecosystem

S. Bijoy Nandan, Asha C.V., Akhilesh Vijay & Jayachandran P.R.

Dept. of Marine Biology, Microbiology & Biochemistry, School of Marine Sciences
Cochin University of Science & Technology, Cochin-682 016, India
[E-mail: bijoynandan@yahoo.co.in]

Introduction

Fjords are semi-enclosed marine inlets that remain under strong terrestrial and oceanic influences and are considered to be strongly dependent on regional species pools of neighboring open seas (Pearson 1980). They are geologically young areas and are regarded as ecologically immature, non-steady state systems that have been evolving over relatively short time scales (Syvitski, 1987). Seasonal or continuous inflows of fresh glacial melt water, loaded with mineral material, influence the structure and circulation of the water masses, turbidity, primary productivity, benthic organisms, sedimentation processes and characteristics. Input of mineral material into the fjord was estimated to be 2.6×10^5 m^3 per season (Elverhoi, Liestol & Nagy, 1980). Sedimentation rate of mineral material in the surface waters was highest close to the glacier front (over 800 g m^{-2} day^{-1}), and then decreases to 200 g m^{-2} day^{-1} at a distance of 4–5 km from the glacier front, and in the outer parts of the fjord does not exceed 25 g m^{-2} day^{-1} (Zajaczkowski, Nygard, Hegseth and Berge, 2010).

The typical feature of the Kongsfjorden system is a well-structured phytobenthic community down to a depth of almost 40 m which plays an important role in primary production, being a food source for herbivores and detritivores, as well as a nursery area and habitat for fish and invertebrates (Hop, Pearson *et al.*, 2002, Lippert, Iken *et al.*, 2001). Shallow water soft bottom macrofaunal communities have only recently received attention but, were not sampled quantitatively. Arctic benthic communities are assumed to have a lower biodiversity compared to other marine systems, especially in the tropical regions and in the Antarctic. This lower biodiversity is thought to be mainly due to the evolutionary history of Arctic Ocean and the uniform habitats dominated by soft bottom organisms, which provide little structure and variety to benthic organisms (Kaczmarek *et al.,* 2005). Benthic communities in arctic fjords are affected by glacial activity in several ways. These includes high concentration and sedimentation rate of mineral suspensions, poorly consolidated and easily re-suspended sediments, fluctuating salinities, low levels of available organic matter, which is diluted by the sedimentation of inorganic material and catastrophic events such as iceberg scouring or sediment slides and slumps (Gorlich *et al.*, 1987, Syvitski *et al.*, 1995, Holme and McIntyre, 2003). In this context, the present study describes the sediment characteristics in relation to the abundance and diversity of benthic fauna in the Kongsforden system.

Materials and methods

Sampling and laboratory analysis

This study was conducted as part of Indian Arctic Expedition at Himadri Station at Spitsbergen, Svalbard, Norway located at the International Arctic Research base, Ny-Ålesund from 17th July 2011 to the 6th August 2011. The Himadri station is operated by National Centre for Antarctic and Ocean Research, Govt. of India. Sampling for the collection of sediment and benthic faunal samples in the Kongsfjorden system, were made on board the research vessel "*Teisten*"of Kings Bay, Svalbard. Seven stations were selected for the summer sampling of various parameters from depths ranging from 76m to 303m (Table 1 & Fig. 1). All samples were collected with a Van Veen grab with a 0.09 m^2 catching area. The samples were sieved through mesh screens with the mesh diameters of 0.5 mm for macrofauna and 63 and 25 µm for meiofauna. The samples were fixed in 4% formalin and in the laboratory all the animals were sorted, identified to the lowest possible taxonomic level and enumerated (Feller, Warwick, 1998, Clarke, Warwick, 1994).

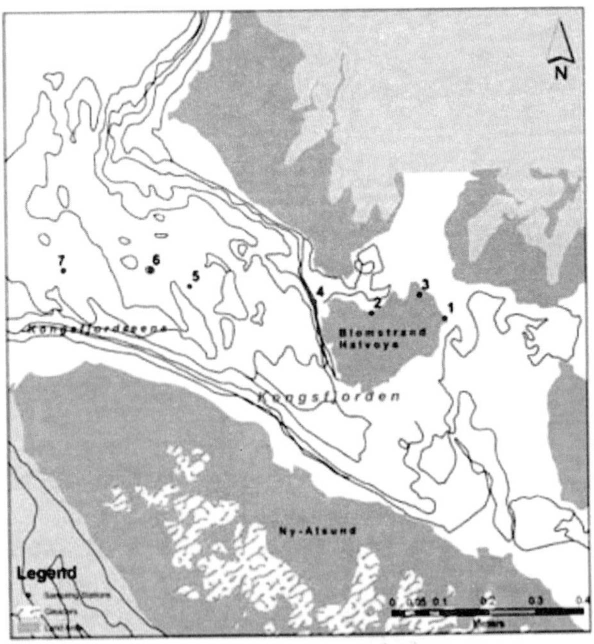

Fig. 1. Map of Kongsfjorden system showing the study stations of the Arctic expedition 2011

Table 1. Details of sampling stations in Kongsfjorden system

Station	Depth	Latitude	Longitude
1	76	78°54.299'	12°13.665'
2	80	78°55.019'	12°04.499'
3	40	78°57.216'	12°10.527'
4	303	78°56.462'	11°57.295'
5	300	78°58.530'	11°41.487'
6	213	79°00.580'	11°36.479'
7	244	79°00.594'	11°25.471'

Based on depth profile, the fjord stations were divided into two subsets representing the outer and inner parts of the fjords. Stations 1, 2 and 3 were considered as inner fjord whereas 4, 5, 6 and 7 as outer. The patterns of species distribution were explored with use of a non-parametric multivariate analysis. The faunal abundance is expressed in percentage abundance and was calculated for each family. *Plymouth Routines in Multivariate Ecological Research (Primer) v6.1* was employed for *Bray-Curtis* similarity index analysis for measuring macro and meiofaunal similarity and species diversity index (Shannon-Wiener log-based index 'H') (Clarke and Gorley, 2006). Sediment granulometric analyses were performed with a Sympatec T 100 laser diffraction granulometer (Sorti and Balsamo, 2010). Sediment pH and oxidation reduction potential; Eh (mv) were determined using standard techniques (Systronics water analyser: NO. MK VI, and No. 318) respectively. The percentage of organic matter content of sample was determined by loss on ignition at 550°C and corresponding organic carbon was estimated using an conversion factor of 1.724 (Alan, Wright, Wang, Reddy, 2008). Inverse distance weighting (IDW) map was made using Arcgis 9.3 to graphically predict the spatial distributions in the different parameters in the study area (Jay Lee, David Wong, 2001).

Results and discussion

Sediments characteristics of the Kongsfjord

In the summer months of August, the average sediment temperature of Kongsfjord ranged from 2.7 to 4.1°C. The organic carbon and Eh of the sediment varied significantly between stations. During the summer months, the high influx of terrigenous material transported by the glacial inputs produces high particle flux, which affected both the benthic food availability and substrate conditions. Sediment in the inner and outer fjord was similar in terms of granulomertic composition. Silt fractions made up 66.3% of the sediment fraction followed by 21% clay and 12% sand (Fig. 2). Granulometric composition of the sediment plays a pivotal role in controlling the macrofaunal densities especially in the highly unstable fjord sediments (Barbara, Aleksandra, 2005). Granulometric characteristics, the quantity of organic carbon and the stability of sediments all reflect the glacial sedimentation gradients in the Arctic glacial fjord (Wlodarska-Kowalczuk *et al.,* 2005). Similarly the salinity and temperature gradients are produced by the inflow of fresh and cold glacial meltwaters, and are controlled by the glacial activity. High levels of mineral sedimentation and sediment deposition have shown to be an acute disturbance agent causing a dramatic decrease in benthic densities and diversity (Airoldi, 2003, Anderson *et. al.,* 2004, Syvitski, Burell *et. al.,* 1987). Sediment pH varied from a lowest value of 6.66 in station 1 to a highest value of 7.39 in station 4. Redox values increased slightly from a lowest value of -141 mv in station 5 to a highest value of -332 mv in station 3 (Fig. 3). IDW prediction map also showed that, the pH increased towards the outer fjord whereas for Eh it decreased towards outer fjord (Fig. 4 & 5). Organic carbon (OC), exhibited values less than 2.5 $g.kg^{-1}$ both in the inner and outer fjord. Surface sediment carbon values varied from a lowest value of 1.0 $g.kg^{-1}$ in station 1 to a highest value of 2.37 $g.kg^{-1}$ in station 5 (Fig. 6). Organic carbon values showed a significant correlation of 0.86 with depth pattern. Organic matter also followed a similar trend that of organic carbon with a lowest value of 1.73 $g.kg^{-1}$ in station 1 to 4.08 $g.kg^{-1}$ in station 5. Similar works from fjords located off Spitsbergen, Greenland, Canadian Arctic, Norway, Scotland and New Zealand has also observed that the organic matter content in surface sediments was lower in the glacial fjords during the summer months than in spring (Wlodarska-Kowalczuk *et al.,* 2004). Decreased concentration of organic matter in the sediment is as a result of glacial mixing. It is predicted through the IDW model that the concentration of organic carbon and organic matter were higher toward the sea, the outer fjord area (Fig. 7).

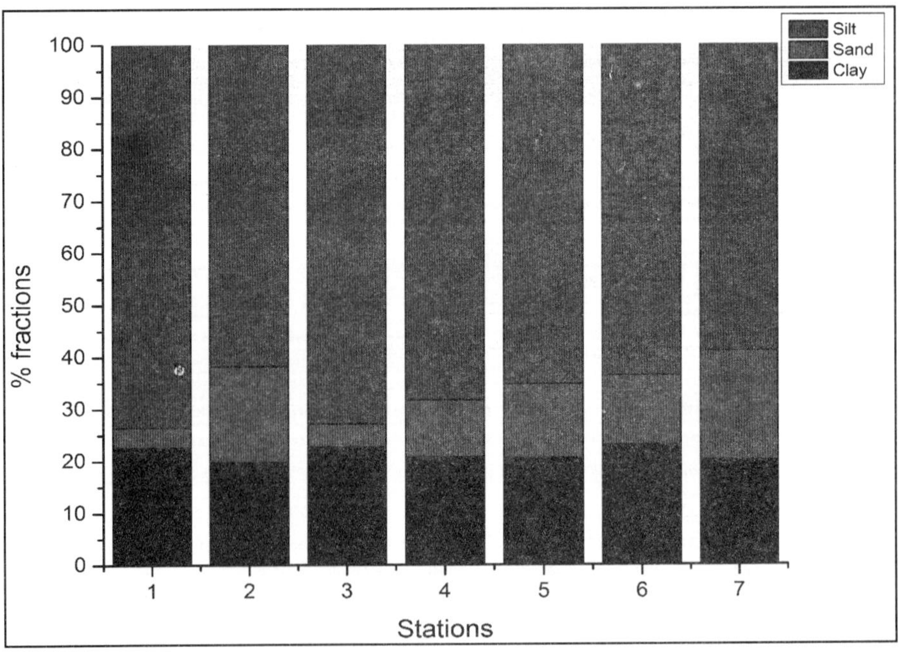

Fig. 2. Granulometric composition of the sediments in the Kongsfjorden system

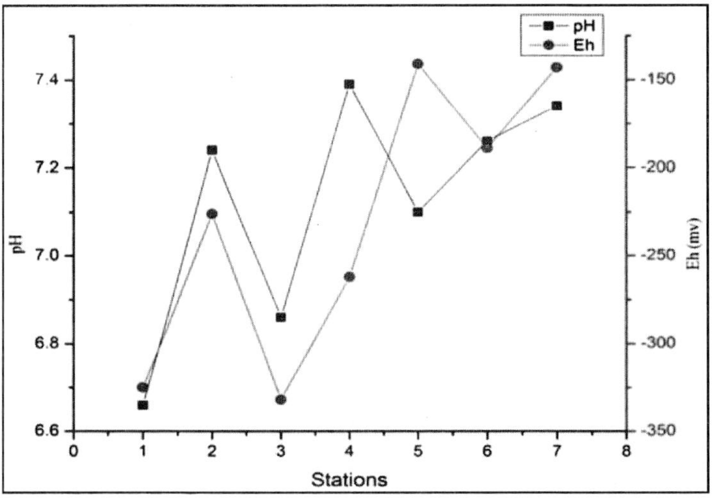

Fig. 3. Station vise variation in pH and EH in the selected stations of Kongsfjord in August 2011

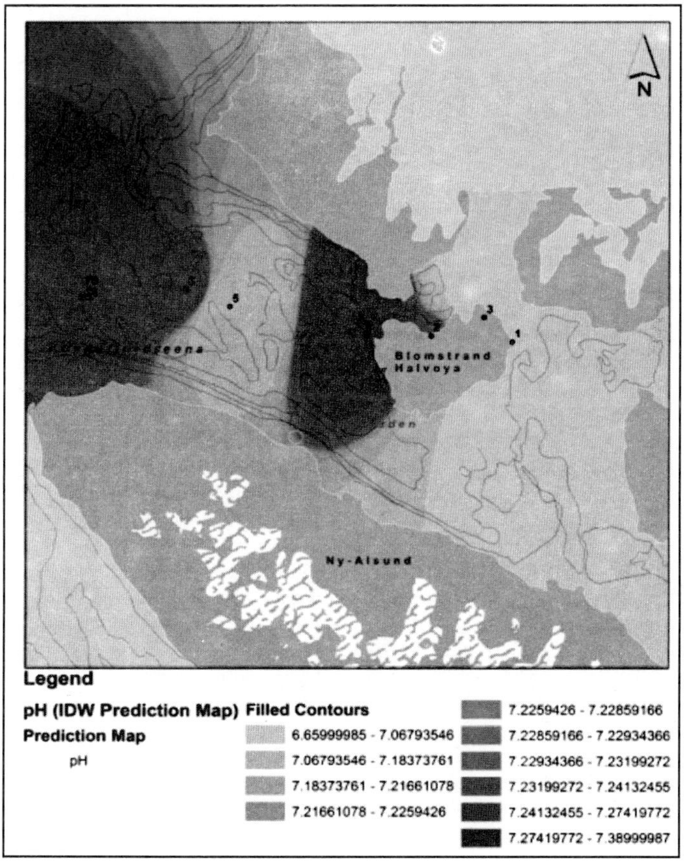

Fig. 4. Inverse Distance Weighting (IDW) prediction map of pH in kongsfjorden system Arctic

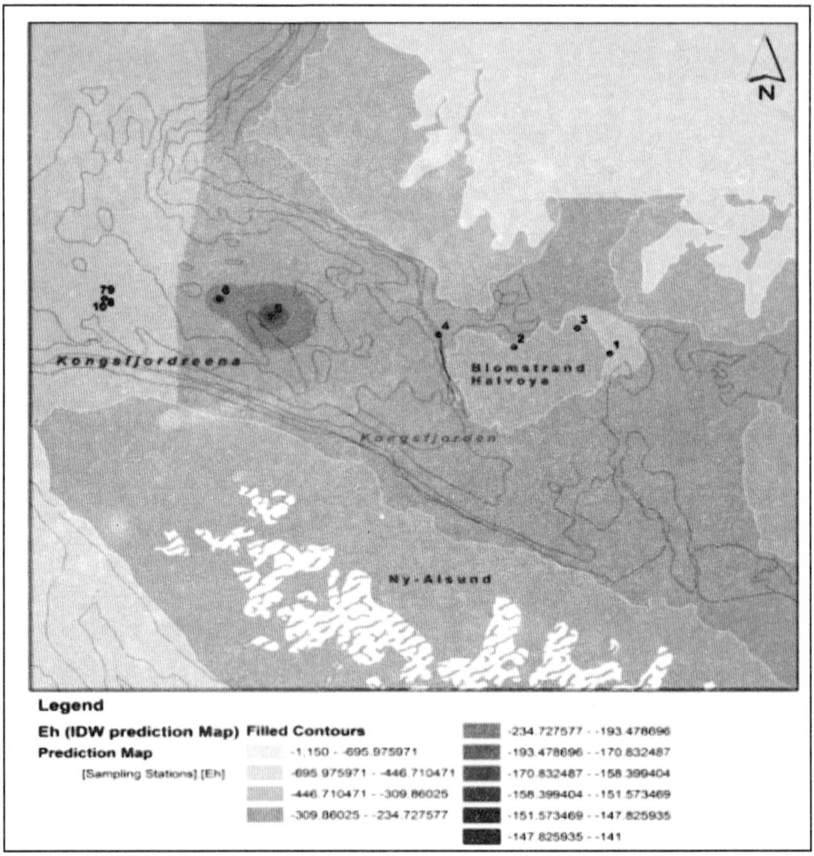

Fig. 5. Inverse Distance Weighting (IDW) prediction map of Eh (mv) in kongsfjorden system Arctic

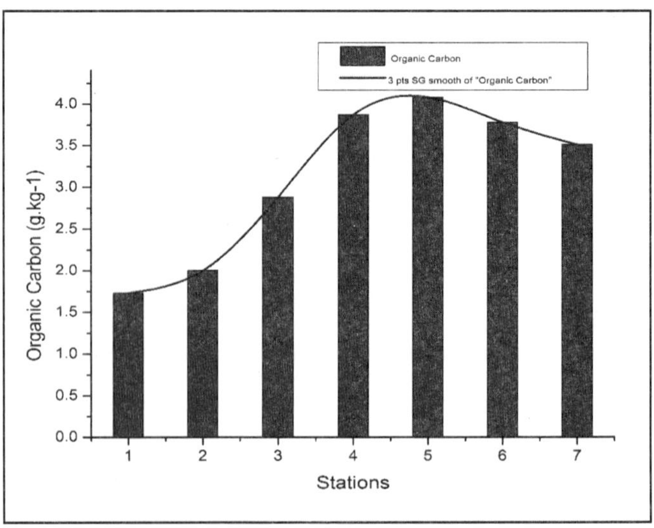

Fig. 6. Organic carbon variation in the selected stations of Kongsfjord in August 2011

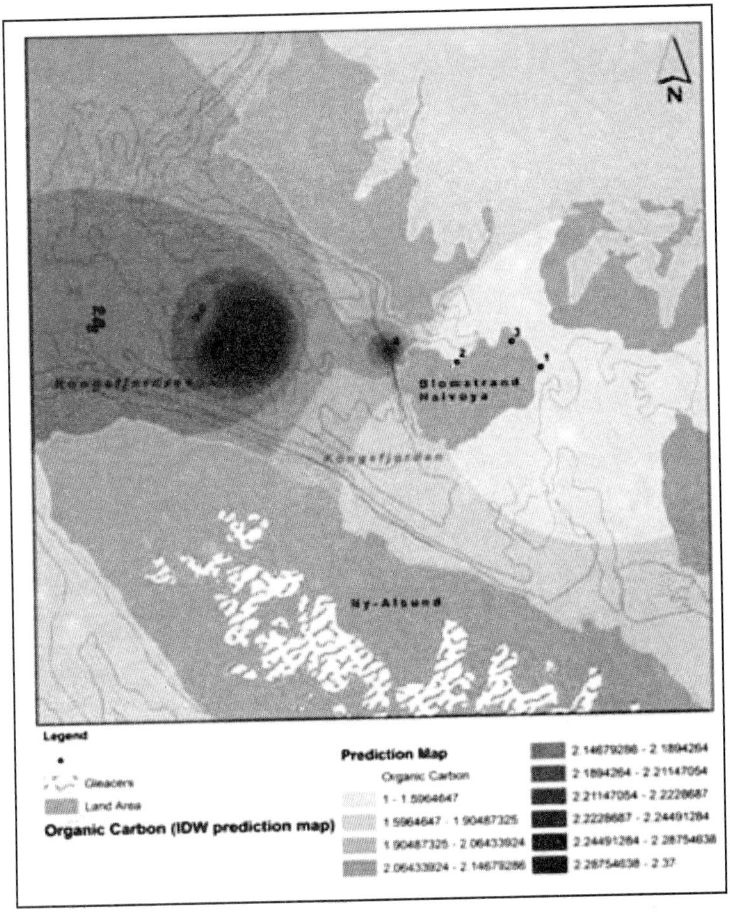

Fig. 7. Inverse Distance Weighting (IDW) prediction map of organic carbon (mg/g) in kongsfjorden system Arctic

Abundance of benthic fauna in Kongsfjord

Twenty one macrobenthic families were identified during the present study in which 13 families were observed in the inner and 9 in the outer fjords. The characteristic species observed from the soft bottom glacial community include foraminiferans, polychaetes, crustaceans and molluscs in which Polychaeta and Foraminifera were represented in highest number. Three polychaete families, Eunicidae (1.39%), Spionidae (0.90%) and Maldanidae (0.57%) were the most abundant in most of the study stations. Legenammine (44.2%), Globobulimina auriculata (27.8%) were the abundant family among foraminiferans (Fig. 8). The inner fjord fauna was dominated by the polycheates family Syllidae, Ampharetidae, *Reophax dentaliniforms* and *Cornuspira folicacea* whereas the outer fjord was constituted by foraminiferans, *Virgulina complanata, Virgulina proboscidea*, Bolivina, *Microchymee acirrata* and Capitellidae. Significant difference in the species richness and diversity of macrobenthic communities in the Kongsfjorden system was observed in the inner and outer fjord region. Shannon Wiener index

(H) also showed a significant variation from a lowest value of 0 in station 6 to of 2.8 in station 2.

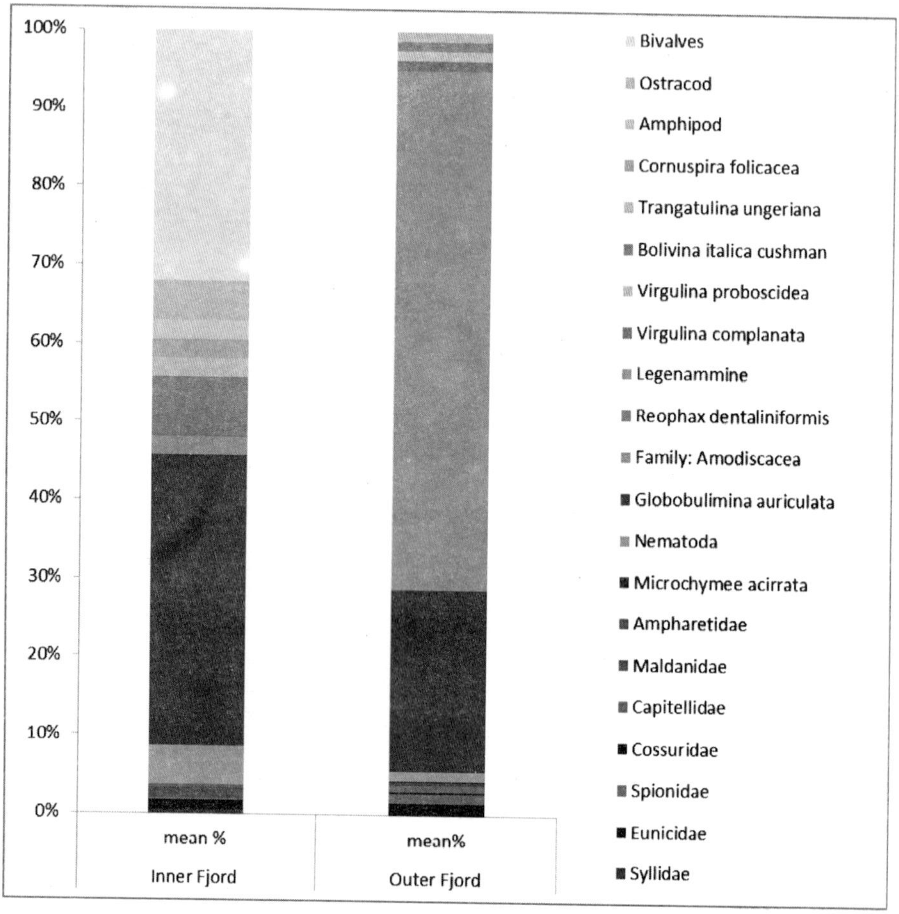

Fig. 8. Average percentage abundance of macrofauna among the sampling locations in the Kongsfjorden system

In the present study, numerical abundance of 33.2% families were recorded from the inner fjord whereas 66.8% from the outer fjord. Patterns of decline in species diversity across the fjord differed on two basic diversity aspects, species richness and evenness. Decline in species richness of macro benthos was observed only in the inner fjord, whereas that of evenness was significantly lower in outer fjord. This pattern has been reported in similar fjords located in different geographical regions and differing in the nature of basic environmental gradients driving the biotic pattern and processes. Significant decrease in the species richness in the inner fjord was documented for all major benthic group, including mostly brooding species (crustaceans) and those with dispersing larva (polycheates and molluscs). These patterns suggest that the connectivity to the open shelf species pool is unlikely to be the main structuring mechanism. Transport of marine species from the offshore shelf environment sustained the initial colonization and is of basic reason for maintenance of species richness in the outer fjord. This can be caused by the difference in the environmental conditions offered by fjord and shelf habitats or competitive exclusion by shelf species (Farrow *et al.*, 1983).

Reason for the decline in faunal abundance and diversity in the Arctic fjord ecosystems are strongly influenced by glacial derived disturbances, mainly by the outflow of meltwater produced by active tidal glaciers. Tidewater glaciers are often situated in the innermost parts of the fjords and they have a strong effect on the physical regimes of whole basins and can impact hard and soft-bottom benthic communities (Syvitski *et al.*, 1987, Thrush *et al.*, 2004). Multivariate analysis of double root transformed macrofaunal abundance formed three clusters with stations 1, 2, 3 and 5 as cluster 1, station 4 and 7 as cluster 2 and station 6 as cluster 3 with an overall similarity of 20% between stations having a stress value of 0.01. Highest similarity was observed between station 2 and 5 (40%) of cluster 1 (Fig. 9). Multivariate analysis distinguishes the inner and outer fjord in terms of macrofaunal abundance. Overall diversity pattern indicated a decline in faunal diversity with increasing depth. Shallow inner fjords showed increased diversity pattern than the deeper outer fjord. Shannon-Wiener diversity indices varied from 1.581 to 3.085 in the inner fjord (station 1 to 3) whereas it was 0 to 3.14 in the outer fjord (station 4 to 7). Organic matter quantity and quality play a critical role in determining macrofaunal distribution in the fjordic system. High-quality organic matter favors the dominance of tube-building, surface-defecators. A higher bulk of organic matter and lower organic quality favor the vertical penetration of infauna and the dominance of subsurface deposit-feeders (Pearson, 1980, Kaczmarek *et al.*, 2005). Natural factors (depth of basins, presence of sills, variable terrestrial runoff, atmospheric forcing, seasonal phytoplankton blooms), and anthropogenic factors (organic enrichment, fishing activities) also leading to community changes, for example, shifts from oxic to anoxic conditions at the sediment surface (Thrush *et al.*, 2004) and seasonal and interannual variability in delivery of organic material to the benthos (Austen *et al.*, 1989, Piepenburg, 2055).

A total of six meiofaunal groups were observed during the summer months of Kongsfjorden glacial system constituted by Foramanifera, Bivalvia, Harpacticoid copepoda, Gastrotricha and Nematoda. The mean percentage abundance of meiofaunal community showed that nematodes (84%) were the dominant group followed by Foraminifera (10.28%), bivalves (1.89%), polycheates (1.80%), harpacticoid copepods (0.25%) and gastrotrichs (0.08%). Maximum numerical abundance of the taxa was recorded in station1 (12,35,772 no.m^{-2}) followed by station 7 (10,03,577 no.m^{-2}) and station 2 (845528 no.m^{-2}) respectively. Nematodes formed the abundant group both in the outer and inner fjord with a percentage abundance of 81.7% in the inner fjord and that of 84.76% in the outer fjord. Foraminifera was constituted by 9.4% in the inner fjord and 12.27% in the outer fjord. The rest of the meiofauna was constituted by bivalves, harpaticoid copepods, gastrotrichs, polycheaetes and invertebrate eggs having an overall percentage of percentage abundance of 8.8% in the inner fjord and 2.9% in the outer fjord (Fig. 10). Multivariate analysis of meiofauna showed an overall similarity of >80% between stations with a stress value of 0.05 (Fig. 11). The MDS plot of meiofauna indicated least variability among stations. Shannon Wiener index (H) varied from a lowest value of 2.0 in station 4 to a highest value of 2.51in station 7. The persistence of stability of meiofaunal abundance and diversity in the inner and outer fjord revealed that meiofauna can be less sensitive than macrofauna to sediment disturbance caused by decreased by sediment re-suspension due to glacial activity.

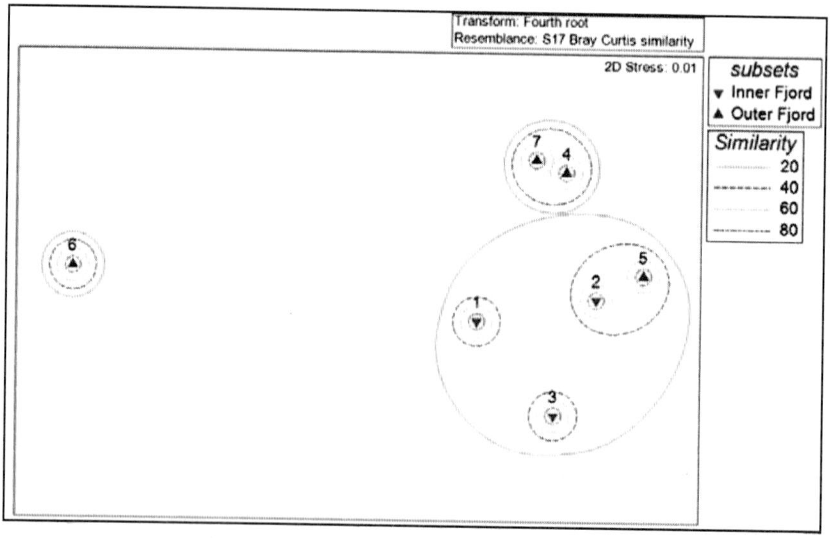

Fig. 9. MDS plot of Bray–Curtis similarities of double-root transformed data of macrofaunal abundance of Kongsfjorden system-Arctic.

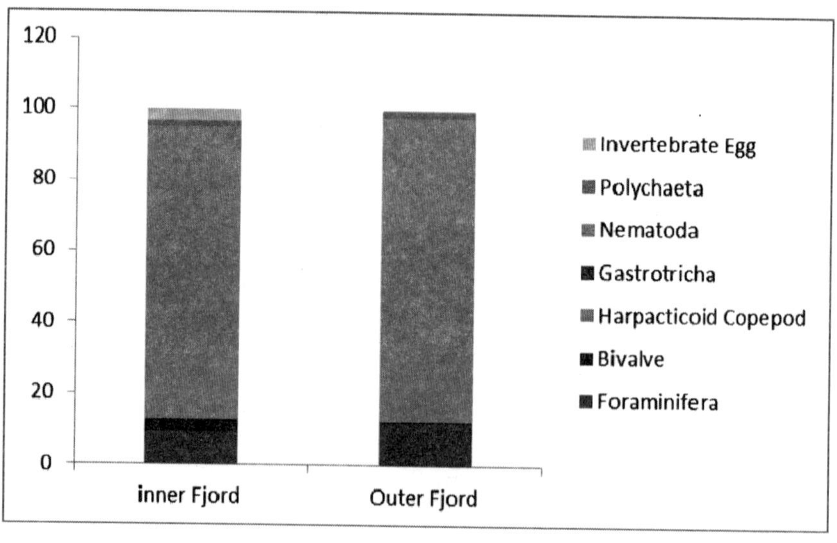

Fig. 10. Percentage abundance of meiofauna among the sampling locations in the Kongsfjorden system

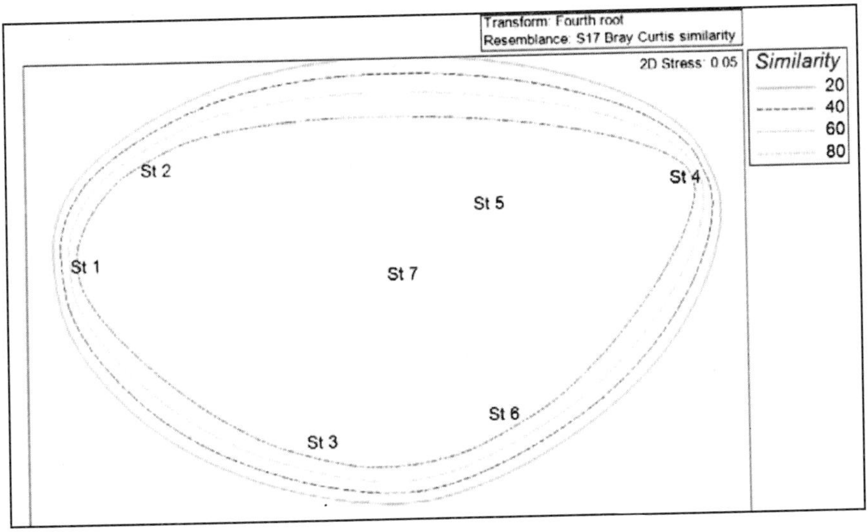

Fig. 11. MDS plot of Bray–Curtis similarities of double-root transformed data of meiofaunal abundance of Kongsfjorden system-Arctic

Conclusion

Based on the preliminary study on the benthic macro and meiofauna conducted in Kongsfjord as part of the Indian Arctic programme has revealed very remarkable observations on the abundance and diversity of the benthic organisms. Declining sediment organic carbon content during the summer months combined with glacial influx had influenced the abundance and diversity of macrobenthic community of the fjord. Diversity and abundance pattern of polychaetes and foraminferan assemblages were greatly influenced by the sedimentary characteristics particularly the organic matter and redox profile. Meiofaunal abundance displayed greater tolerance to the sediment quality as compared to the macrofauna in relation to the glacial activity. Present study corroborates to some of the earlier works conducted in the system. Present study could also be considered a harbinger on the influence of the global climate change impacts in the arctic represented by Kongsfjorden system.

Acknowledgements

Authors are thankful to the National Centre for Antarctic and Ocean Research (NCAOR), Govt. of India and to the Head, Dept. of Marine Biology, Microbiology and Biochemistry, School of Marine Sciences, Cochin University of Science and Technology for providing the support and necessary facilities to undertake the study.

References

Airoldi L, The effects of sedimentation on rocky coast assemblages, Oceanogr. Mar. Biol. Ann. Rev., 41(2003) 161–236

Alan L, Wright Y, U Wang & K R Reddy, Loss- on- ignition method to assess soil organic carbon in calcareous everglades wetlands, Soil science and Plant Analysis, 39(2008) 3074- 3083

Anderson M J, Ford R B, Feary D A & Honeywill C, Quantitative measures of sedimentation in an estuarine system, and its relationship with intertidal soft-sediment infauna, Mar. Ecol. Prog. Ser., 272(2004) 33–48

Austen M C, Warwick R M & Rosado M C, Meiobenthic and Macrobenthic structure along a putative pollution gradient in southern Portugal, Mar Pollut., 20(1989) 398-405.

Barbara U E J W & Aleksandra J T M, Intertidal meiofauna of a high-latitude glacial Arctic fjord (Kongsfjorden, Svalbard) with emphasis on the structure of free-living nematode communities, Polar Biol., 28(2005) 940–950

Clarke K R & Gorley R N, Primer v6: User Manual/Tutorial. (PRIMER-E: Plymouth, UK) 2006, pp. 9

Clarke K R & Warwick R M., Changes in marine communities: an approach to statistical analysis and interpretation, (Natural Environment Research Council, Plymouth Marine Laboratory, Plymouth) 1994

Elverhoi A, Liestol O & Nagy J, Glacial erosion, sedimentation and microfauna in the inner part of Kongs-fjorden, Spitsbergen. Geological and geophysical research in Svalbard and on Jan Mayen., 172 (1980), 33–60

Farrow G E, Syvitski J P M & Tunnicliffe V, Suspended particulate loading on the macrobenthos in a highly turbid fjord: Knight Inlet, British Columbia, Can, J Fish Aquat Sci., 40 (1983) 273– 288

Feller R J & Warwick R M, Energetics. Introduction to the study of meiofauna, (In:Higgins RP, Thiel H (eds), Smithsonian Institution press, Washington) 1998, pp. 181-196

Gorlich K, Weslawski J M & Zajaczkowski M, Suspension settling effect on macrobenthos biomass distribution in the Hornsund fjord, Spitsbergen, Polar Res., 5(1987) 175–192

Holme N A & Mc Intyre A D. Methods for study of Marine Benthos, (IBP Hand book No.6, Blackwell Scientific Publications) 2003

Hop H, Pearson T H, Hegseth E N, Kovacs K M, Wiencke C, Kwasniewski S, Eiane K, Mehlum F, Gulliksen B, Wlodarska-Kowalczuk M, Lydersen C, Weslawski J M, Cochrane S, Gabrielsen G W, Leakey R, Lonne O J, Zajaczkowski M, Falk-Petersen S, Kendall M A, Wängberg S A, Bischof K, Voronkov A, Kovaltchouk N A, Wiktor J, Poltermann M, di Prisco G, Papucci C & Gerland S, The marine ecosystem of Kongsfjorden, Svalbard, Polar Resh., 21(2002) 167–208

Jay Lee, David W & S Wong, Statistical Analysis with Arcview Gis (John wiley & sons united states of America) 2001

Kaczmarek H, Wlodarska Kowalczuk M, Legezynska J & Zajaczkowski M, Shallow sublittoral macrozoobenthos in Kongsfjord, West Spitsbergen, Svalbardm, Po.l Polar Res., 26 (2005) 137–155

Lippert H, Iken K, Rachor E & Wiencke C, Macrofauna associated with macroalgae in the Kongsfjord (Spitsbergen), Polar Biol., 24(2001) 512-522

Pearson T H, Macrobenthos of fjords, (Freeland HJ, Farmer DM, Levings CD (eds) Fjord oceanography NATO Conference. Plenum, New York) 1980 pp. 569–602

Piepenburg D, Recent research on Arctic benthos: common notions need to be revised, Polar Biol., 28 (2005) 733–755

Sorti F & Balsamo F, Particle size distribution by laser diffraction: sensitivity of granular matter strength to analytical operating procedures, Solid Earth., 1(2010) 25-48

Syvitski J M P & Shaw J, Sedimentology and geomorphology of fjords. (In: Perillo GME (ed) Geomorphology and Sedimentology of estuaries. Developments in sedimentology 53. Elsevier Sci B.V, Amsterdam) 1995, pp.113–178

Syvitski J P M, Burell C D & Skei J M, Fjords. Processes and products (Springer, Berlin Heidelberg New York (1987)

Thrush S F, Hewitt J E, Cummings V J, Ellis J I, Hatton C, Lohrer A & Norkko A, Muddy waters: elevating sediment input to coastal and estuarine habitats, Frontiers in Ecology and Environment, 2 (2004) 299–306

Wlodarska Kowalczuk M & Pearson T H, Soft-bottom macrobenthic faunal associations and factors affecting species distributions in an Arctic glacial fjord (Kongsfjord, Spitsbergen), Polar Biol., 27(2004), 155–167

Wlodarska-Kowalczuk M, Pearson T H & Kendall M A, Benthic response to chronic natural physical disturbance by glacial sedimentation in an Arctic fjord, Mar. Ecol. Prog. Ser., 303 (2005) 31–41.

Zajaczkowski M, Nygard H, Hegseth E N & Berge J, Vertical flux of particulate matter in an Arctic fjord : the case of lack of the sea-ice cover, Adventfjorden 2006-2007, Polar Biol., 33(2010) 223-239.

Subject Index

Other Books

Climate Change & Himalayan Ecosystem-indicator, Bio & Water Resources

J. Sundaresan, Pankaj Gupta, K.M. Santosh & Ram Boojh (Eds.)

"Climate Change and Himalayan ecosystem-indicator, Bio and water resources" consists basic information and data on glaciers, climate change indicators & projections, water resources and bio-diversity hot spots of Mount Himalaya. Studies on "Climate change and the recession pattern of the Glaciers in the Himalaya" of this book concludes that "Possibility of the rivers in the Himalayas drying up as a consequence of rapid degeneration of the glaciers is not borne out by the past history". In this book, study conducted in watershed of Central Himalaya, a Decision Support System (DSS) is introduced as interactive tool that understands the problem and explores various courses about water demand and supply management to help decision makers. Himalayan foreland basin derivatives hold records of climatic changes in response to monsoonal circulation. In this study detrital records of Ramganga sub-basin of HFB are focused to understand the climate aspect during its deposition. Himalayan biodiversity conservation is discussed in detail in this book. It infers that in Himalaya with the current technological capability, it is very certain that the present species extinction rate will overtake the biodiversity inventorization and characterization. Carbon sequestration potential of the forests of Himalaya is analyzed in this book. This book has a detailed account of the altitudinal shiftiness of butterflies due to increase of air temperature at West Kameng district of Arunachal Pradesh. Changes in NDVI (Normalized Difference Vegetation Index) over a period of several years, is examined in this study to assess the changes caused by climate or socioeconomic aspects. This book will be a hand book for researchers, students, environmentalist and to administrators who are associated with multi dimensional decision support system in Mountain ecosystem.

ISBN: 978-81-7233-847-3, 2013, 216 + xiv pages

Other Related Books

Climate Change Adaptation Strategies in Agriculture and Allied Sectors
GSLHV Prasad Rao

Climate Change and Environment
J. Sundaresan et al.

Climate Change Impact on Ecosystem
J. Sundaresan et al.

Climate Change and its Ecological Implications for the Western Himalaya
V.L. Chopra